Future Curricular Trends in School Algebra and Geometry

Proceedings of a Conference

A Volume in
Research in Mathematics Education

Series Editor:
Barbara Dougherty

Research in Mathematics Education

Barbara Dougherty
Series Editor

The Classification of Quadrilaterals: A Study in Definition (2008)
By Zalman Usiskin

A Decade of Middle School Mathematics Curriculum Implementation: Lessons Learned from the Show-Me Project (2008)
Edited by Margaret R. Meyer and Cynthia W. Langrall

The History of the Geometry Curriculum in the United States (2008)
By Nathalie Sinclair

Mathematics Curriculum in Pacific Rim Countries—China, Japan, Korea, and Singapore: Proceedings of a Conference (2008)
Edited by Zalman Usiskin and Edwin Willmore

The Intended Mathematics Curriculum as Represented in State-Level Curriculum Standards: Consensus or Confusion? (2006)
Edited by Barbara Reys

MR

Future Curricular Trends in School Algebra and Geometry

Proceedings of a Conference

Edited by
Zalman Usiskin
Kathleen Andersen
Nicole Zotto
The University of Chicago

INFORMATION AGE PUBLISHING, INC.
Charlotte, NC • www.infoagepub.com

Library of Congress Cataloging-in-Publication Data
International Mathematics Curriculum Conference (2nd : 2008 : Chicago, Ill.)
Future curricular trends in school algebra and geometry : proceedings of a conference / [edited by] Zalman Usiskin, Kathleen Andersen, Nicole Zotto.
p. cm. – (Research in mathematics education)
"Proceedings of the Second International Curriculum Conference sponsored by the Center for the Study of Mathematics Curriculum (CSMC), held May 2-4, 2008 at The Field Museum and on the campus of the University of Chicago"–Pref.
Includes bibliographical references.
ISBN 978-1-61735-006-1 (pbk.) – ISBN 978-1-61735-007-8 (hardcover) – ISBN 978-1-61735-008-5 (e-book)
1. Algebra–Study and teaching (Elementary)–Curricula–Congresses. 2. Geometry–Study and teaching (Elementary)–Curricula–Congresses. 3. Arithmetical algebraic geometry–Study and teaching (Elementary)–Curricula–Congresses. I. Usiskin, Zalman. II. Andersen, Kathleen. III. Zotto, Nicole. IV. Title.
QA159.I58 2008
372.7–dc22

2010006705

Printed in the United States of America

CONTENTS

Preface .... **vii**

1. Introductions **1**

PART I
EARLY ALGEBRA

2. Early Algebraic Thinking: The Case of Equivalence in an Early Algebraic Context **7**
Elizabeth Warren

3. A Brief Essay on the Need to Consider the "Superficial" Aspects of Learning Algebra **31**
Romulo Lins

4. Early Algebra **45**
Maria L. Blanton

5. A Davydov Approach to Early Mathematics **63**
Barbara J. Dougherty

PART II
ALGEBRA USING COMPUTER ALGEBRA SYSTEMS (CAS)

6. Technology and the Yin and Yang of Teaching and Learning Mathematics **73**
Bernhard Kutzler

7. **CAS and the Future of the Algebra Curriculum........................... 93**
Kaye Stacey

8. **Algebra in the Age of CAS: Implications for the High School Curriculum: Examples from the CME Project............................109**
Al Cuoco

9. **A Perspective on the Future of Computer Algebra Systems in School Algebra..129**
M. Kathleen Heid

PART III

3-D GEOMETRY

10. **Three-Dimensional Citizens Do Not Deserve a Flatlanders' Education: Curriculum and 3-D Geometry.................................147**
Claudi Alsina

11. **Manipulating 3D Objects in a Computer Environment..............155**
Jean-Marie Laborde

12. **Algebra and Geometry: From Two to Three Dimensions...........169**
Thomas F. Banchoff

13. **Thoughts on Elementary Students' Reasoning about 3-D Arrays of Cubes and Polyhedra 183**
Michael T. Battista

PART IV

LINKING ALGEGBRA AND GEORMETRY

14. **Linking Geometry and Algebra in the School Mathematics Curriculum 203**
Keith Jones

15. **Linking Geometry and Algebra through Dynamic and Interactive Geometry...217**
Colette Laborde

16. **Linking Algebra and Geometry: The Dynamic Geometry Perspective 231**
Nicholas Jackiw

17. **Linking Algebra and Geometry in the Interactive Mathematics Program 243**
Diane Resek

PART V
PANEL OF PRACTITIONERS AND CURRICULUM DEVELOPERS

18. **Making Future Trends Realities In US Classrooms 253**
Diane J. Briars

19. **Tools, Technologies, and Trajectories 259**
Douglas H. Clements

20. **Future Trends in School Algebra and Geometry: Reflections on the Vision of Experts 267**
James Fey

21. **Thoughts From a Classroom Teacher 273**
Jim Mamer

22. **Restoring and Balancing 277**
William McCallum

PART VI
REFLECTIONS BY CONFERENCE ATTENDEES

23. **Insights into Dynamic Mathematical Learning Environments.... 287**
Sarah J. Hicks, Melissa D. McNaught, and J. Matt Switzer

24. **Instrumental Genesis and Future Research in School Algebra and Geometry 295**
Daniel J. Ross

25. **Closing Remarks: Reflections from a Retiring Mathematics Curriculum Developer 305**
Zalman Usiskin

APPENDICES

Conference Schedule 313

Biographies of Presenters 318

Attendees 329

PREFACE

This volume contains the proceedings of the Second International Curriculum Conference sponsored by the Center for the Study of Mathematics Curriculum (CSMC), held May 2–4, 2008 at The Field Museum and on the campus of the University of Chicago.

The CSMC is one of the National Science Foundation Centers for Learning and Teaching (Award No. ESI-0333879). As noted on the CSMC website (http://mathcurriculumcenter.org), the CSMC serves the K–12 educational community by focusing scholarly inquiry and professional development around issues of mathematics curriculum. Major areas of work include understanding the influence and potential of mathematics curriculum materials, enabling teacher learning through curriculum material investigation and implementation, and building capacity for developing, implementing, and studying the impact of mathematics curriculum materials. The work of the CSMC is not driven by a particular philosophy or ideology. As researchers, CSMC staff try to maintain a healthy skepticism throughout all phases of the center's activities considering both international and multiple U. S. perspectives with an ultimate goal to produce research-based knowledge and products that will enlighten and serve the range of users of mathematics curriculum materials.

The venue, the University of Chicago, was also the location of the first CSMC International Conference held in November, 2005. This conference differed from the first in that it focused on content of the curriculum,

Future Curricular Trends in School Algebra and Geometry: Proceedings of a Conference. pages ix–x

whereas the first conference had focused more on the process of creating curriculum. The earlier conference, whose proceedings are also published by Information Age Publishing, was directed at learning about recent trends in four carefully chosen countries, as its title "Mathematics Curriculum in Pacific Rim Countries—China, Japan, Korea, and Singapore"—suggests.

The success of the first conference led us to structure the second conference in a way analogous to the first. Instead of four countries, we chose four topics—early algebra, computer algebra systems (CAS), 3-dimensional geometry, and linking algebra and geometry. We invited two speakers from other countries on each topic, thus eight speakers in all. We asked these speakers to provide insights about implementing these ideas.

Then, complementing these presentations, we invited two researchers from the United States whose work involved these four topics, again giving us a total of 16 speakers. The conference ended with two sessions reacting to the earlier sessions. The first session was a panel of four practitioners and curriculum developers speaking about the practical problems of implementing ideas like the ones that were promulgated at the conference. The second session included reactions from two national leaders.

In this volume are papers from all of the speakers at the conference except one.

Before the conference, doctoral students from the CSMC home universities and several other institutions were informed that they would be asked to write a paper reacting to the conference. We invited the students to write the paper either singly or in groups and indicated that we would pick the best of these papers for publication. Two papers were written and we felt that both were worthy, so both are included in this volume.

ACKNOWLEDGEMENTS

A conference of this type requires the work of many people and the support of many organizations.

We thank first the National Science Foundation, without whose support this conference could not have taken place.

We thank the many speakers from the U.S. and abroad, who shared their thoughts and expertise with us, and the conference participants who contributed to the conversation and ambience of the meeting.

This volume was edited by Kathleen Andersen, Nicole Zotto, and Zalman Usiskin. We wish also to thank Barbara Dougherty for her role in bringing these proceedings to publication, and to Information Age Publishing for publishing the manuscript.

CHAPTER 1

INTRODUCTIONS

The opening session of the conference took place in the evening, in the James Simpson Theater of The Field Museum. The chair of the conference planning committee and the Director of the Center for the Study of Mathematics Curriculum (CSMC) welcomed the nearly 200 attendees. The following is a transcription of the welcoming remarks.

Zalman Usiskin

I am Zalman Usiskin, chair of the planning committee, and it is my honor and pleasure today to welcome you to Chicago on behalf of that committee and the CSMC.

There are many reasons to hold international conferences. One reason is that we always learn things at these conferences that we did not know before. We learn that some problems that we thought were unique to the U.S. are general worldwide. We learn that some solutions that we might consider controversial are being used in other countries or mired in controversies in those countries. We gain from the different perspectives that people in other countries have.

Another reason for having these conferences is to display the complexity of curricular issues that are known to all of us who do curriculum. We have

Future Curricular Trends in School Algebra and Geometry: Proceedings of a Conference. pages 1–4

long seen the too-quick acceptance of the simplistic argument that goes as follows: Someone in country B says that country A is doing well—so let's copy the curriculum of country A. But when we speak to people in country A, almost invariably they say that we are speaking of "the old curriculum" and they have already changed their curriculum. So what are we copying?

We also learn that people in other countries are interested in what we do. Policy makers and researchers trying to figure out what goes on here are often frustrated because it is so difficult to determine what is going on in our country. We have so many different curricula in use or being tried out, but that weakness is our strength. We know that some of the best work in the world in mathematics curriculum is in the U.S., and people in other countries look to this work to improve their curricula.

These are general reasons for holding a curriculum conference. But, specifically, why this second CSMC conference? The first CSMC conference looked at the present: how Asian countries, known for high performance, decide on what they will teach and how they move from their national curricula to textbooks.

We felt an obligation in the second CSMC conference to look at the future: what current mathematics curriculum research in other countries might have the most implications for research and thinking about curriculum in the United States. We thought about focusing either on algebra or on geometry but decided to include both because so many connections exist between them. As in the first conference, we have invited people in the United States who are researching the same topics as our international speakers, both to showcase the work of U.S. scholars and to help ideas from other countries to reach relevant people, and ideas from the U.S. to be spread outside our borders.

We picked four areas of curriculum research in algebra and geometry known to be of interest both here and abroad: early algebra, CAS, 3-D geometry, and links between algebra and geometry.

I am pleased to report that, after too long a time, the proceedings of the first conference should be available in about a month from Information Age Publishing, and we hope that the proceedings of the second conference will be published in the same manner. But you will not necessarily have to wait for proceedings. We have asked the speakers if we can place their powerpoints online as soon as possible, and if you go to the CSMC website even before the end of the conference you may be able to see some of them.

Now it gives me great pleasure to introduce Barbara Reys, one of the CSMC co-directors, and the person who keeps all of us associated with CSMC on task.

Barbara Reys

On behalf of the other co-directors of CMSC—Glenda Lappan and Chris Hirsch—I'd like to welcome those of you who have traveled near and far to this conference. In particular, I want to welcome and thank our international guests from Australia, Austria, Brazil, Canada, England, France, and Spain for participating.

The Center for the Study of Mathematics Curriculum serves the K-12 educational community by focusing scholarly inquiry and professional development around issues of mathematics curriculum.

An important part of the work of CSMC is to stimulate discussions, nationally and internationally, focused on improving student learning opportunities through the articulation and development of high quality curriculum standards and materials. We have much to learn from each other, and I encourage participants to engage with and challenge each other so that we leave with renewed enthusiasm and new ideas to move our collective work forward.

I want to take this opportunity to express my sincere thanks to Zalman Usiskin and his associates, Carol Siegel and Kathleen Andersen, at the University of Chicago for the hard work of organizing and hosting this conference. They've attended to countless details that will ensure an organized and productive meeting.

Finally, I ask that those associated with CSMC stand for a few moments. These folks will have a "blue" nametag so please feel free to ask them for help if you need it and for information about the work of the Center. Please enjoy the conference.

A second opportunity for welcoming the participants occurred the next morning of the conference, when the venue moved to the University of Chicago. There a welcome came from the President of the University.

Zalman Usiskin

I would like to read a welcome from Robert Zimmer, the President of the University. Bob was planning to be here in person, but he has an out-of-town engagement that has made this impossible. It is disappointing because he is a Professor of Mathematics, an ex-chair of the mathematics department here, and a person who is very knowledgeable about the work of the University of Chicago School Mathematics Project.

Robert J. Zimmer

I would like to welcome all of you to the International Mathematics Curriculum Conference. The University of Chicago is pleased to join with The Field Museum in hosting the conference and proud to be one of the four universities who are partners in the Center for the Study of Mathematics Curriculum.

Mathematics has had a long and distinguished tradition at the University of Chicago since the founding of the University in 1892. This tradition has been manifest not only in research contributions to advanced mathematics, but to decades of attention, commitment, and achievement in the area of school mathematics curriculum. This is a vitally important area, both because it allows students to achieve their full personal potential, and because of the societal importance of a mathematically educated citizenship. This weekend's conference will continue the tradition of people coming to Chicago from around the world to discuss mathematics curriculum and share their ideas about teaching and learning.

I have a long personal connection to and interest in school mathematics education, and I had hoped to be able to welcome you to our campus and to this conference in person. Unfortunately, I needed to be out of town, so I hope you accept this welcome in absentia with the enthusiasm with which it is conveyed.

I wish you all the best with the conference. Please enjoy the day and the campus.

Robert J. Zimmer, President
The University of Chicago

For your reference, a conference program can be found on pages 313–318.

PART I

EARLY ALGEBRA

CHAPTER 2

EARLY ALGEBRAIC THINKING

The Case of Equivalence in an Early Algebraic Context

Elizabeth Warren

INTRODUCTION

In 2002 I became part of a group who were developing the new Mathematics syllabus for Queensland. Knowing the difficulties that many secondary students experienced with algebra, I was particularly interested in developing a Patterns and Algebra strand for this syllabus. Together with a colleague, Tom Cooper, we developed a draft proposal for the Patterns and Algebra strand for consideration. Simultaneously, Professor Cooper and I began a longitudinal study with five elementary schools in Queensland. The timing of this study, the Early Algebraic Thinking Project (EATP)[1] could not have been better. It gave us the opportunity to work with classroom teachers to ascertain:

a. whether elementary students could engage in algebraic thinking,
b. if they could, what representations and teacher actions best support and enhance this engagement; and

[1]EATP was funded by the Australian Research Council Linkage grant LP0348820

Future Curricular Trends in School Algebra and Geometry: Proceedings of a Conference. pages 7–30

c. what path do students follow as they engage in this learning.

The literature presents two differing perspectives on the ontology of student learning namely, the learning trajectory and the learning-teaching trajectory. While both perspectives have many commonalities, the main differences lie in their emphasis on the act of teaching in the learning process, and their prescriptiveness of the resulting curriculum. From the first perspective learning consists of a series of 'natural' developmental progressions identified in empirically-based models of children's thinking and learning (Clements, 2007). In conjunction with viewing learning as a progression through development hierarchical levels, the learning trajectory sees teaching as the implementation of 'a set of instructional tasks designed to engender these mental processes' (Clements, 2007). From this perspective the act of teaching is secondary to the act of learning. The resulting curriculum consists of diagnostics tests, learning hierarchies and purposely-selected instructional tasks. Fundamental to this perspective is (a) the existence of a large repertoire of empirically-based research evidencing the development of particular concepts, such as, number, number sense and counting, and (b) conducting extensive field tests trialing various instructional tasks.

In contrast, the learning-teaching trajectory has three interwoven meanings each of equal importance. These are: a learning trajectory that gives an overview of the learning process of students; a teaching trajectory that describes how teaching can most effectively link up with and stimulate the learning process; and, finally, a subject matter outline, indicating which core elements of the mathematical curriculum should be taught (Van den Heuvel-Panhuizen, 2008). It is believed that the learning-teaching trajectory provides a 'mental education map,' which can help teachers to make didactical decisions as they interact with students' learning and instructional tasks. It serves as a guide at the meta level. The resulting curriculum tends to be more open and flexible with teachers choosing and adapting activities in order to enhance student learning. Learning is not necessarily seen as a progression through hierarchical steps. It is this second perspective that has greatest resonance with the research presented in this paper.

Early algebraic thinking is a relatively new area in the elementary curriculum. Thus current research in early algebra tends to be of an exploratory nature, with a focus on explicating key transitions in children's learning and teaching activities that support crossing these transitions. We also believe that on a day-to-day basis, the act of teaching is as important as the act of learning. Teachers make decisions about the selection of tasks and representations used in the tasks. Learning is not necessarily a step-by-step process progressing through hierarchical levels. Thus our aim was not only to begin to identify key transitions in student learning in the domain of early algebra, but also to identify particular teaching actions that supported these transitions. We were particularly interested in how different representations assisted this teaching and learning dance.

We developed a framework for early algebra based on our knowledge and beliefs at the time, which were informed by the structural view of mathematics (Sfard, 1991), a cognitive perspective on learning (English & Halford, 1995; Hiebert & Carpenter, 1992), and an appreciation of students' difficulties with variables and the cognitive gap between arithmetic and algebra (Linchevski & Herscovics, 1996; Usiskin, 1988). This framework formed the basis of the new Queensland mathematics syllabus. One of the problems that occur worldwide with secondary students and their transition from arithmetic to algebra is their lack of an appropriate arithmetic knowledge base suitable for building algebraic concepts at the completion of elementary school.

Presently, many young children (even at the end of Year 2) already possess understandings of arithmetic that are very narrow and impede the development of algebraic thinking (e.g., Carpenter, Franke, & Levi, 2003). Once these misconceptions exist they are very hard to change (Carpenter et al., 2003) and become even more entrenched as these children progress through schooling (e.g., Warren, 2003). Thus our aim was to 'open up' discussions about arithmetic in the elementary classroom with a refocus on the properties and principles of mathematics rather than the continual focus on finding answers to arithmetic problems.

PERSPECTIVE ON MATHEMATICS OF EARLY ALGEBRA

In its most powerful form, we viewed algebra as an abstract system, the interactions which reflected the structure of arithmetic (Usiskin, 1988). Our stance reflects that of Dienes (1961) and Skemp (1978) who see the importance of algebra as the way it represents principles (such as the commutative law and the balance principle) and the structures of mathematics (such as the field properties and equivalence class). We did not consider aspects such as the factorisation and simplification as the essence of algebra, but rather algebra as a system characterised by indeterminacy of objects, an analytic nature of thinking and symbolic ways of designating objects (Radford, 2006).

In our conception of algebra we considered it consisting of two core approaches, relations and change. These reflect the foci of Scandura (1971), namely, things, relationships between things and the transformations (changes) between things. Thus operations can be seen as either relational or static (e.g., $3 + 4 = 7$, three and four is seven) or as transformational or changing (e.g., $3 + 4 = 7$, three changes by 4 to give seven). From a relational perspective equals is equivalence or 'same value as' but from a transformational perspective equal is seen as two way mapping, one side to the other (Linchevski, 1995). This perspective mirrors two of Kaput's (2006) three core strands: algebra as a study of structure and systems abstracted from computations and relations; and, algebra a study of functions, relations and joint variation. Malara and Navarra (2003) make a clear distinction between algebraic thinking and arithmetic thinking in an elementary context: the former focuses on process where as the second focuses on product. This distinction assisted us to delineate between the two in classroom discussions, and to move from one to the other as the need arose. At times we needed arithmetic to support algebraic thinking (e.g., generalising the compensation principle of arithmetic) while at other times we needed algebraic thinking to support arithmetic (e.g., adding 3 to 2 is the same process as adding 3 to 82, 3 to 1012, 30 to 20). We concur with Mason's (2006) claim that the power of mathematics lies in the intertwining of algebraic thinking and arithmetic thinking, each enhances the other as students become numerate.

The framework that we developed for EATP encompassed: (1) patterns and functions, a study of repeating and growing patterns and of early functional thinking, (2) equivalence and equations, a study of equivalence, equations and expressions, and (3) arithmetic generalisations, a study of number that involves generalisation to principles. The framework also saw early algebra as the development of mental models on the basis of relationships between real world instances, symbols, language, drawings and graphs, particularly those that enabled the modelling of real situations that contained unknowns and variables.

This paper reports on one aspect of EATP, students' ability to generalise arithmetic structure, particularly to comprehend expressions and equations (the second component of our framework), mathematics as relationships rather than mathematics as change. It is a sweep across 5 years with a cohort of Year 2 to 6 students and describes the journey that occurred as we traversed the territory. There is particular emphasis on impediments to developing this thinking and the representations, discussions and actions that began to allow students to engage with equivalent ideas and to use this understanding in solving problems with unknowns. Fundamental to these discussions were the interweaving of dialogue within a representational system.

THE ROLE OF REPRESENTATIONS

There has been general consensus that mathematical ideas are presented externally (e.g., concrete materials, pictures/diagrams, spoken words, written symbols) and internally (mental models, cognitive representations) and that mathematical understanding is exhibited by the number and strength of connections in the students' internal network of representations (e.g., Hiebert & Carpenter, 1992). In this perspective the development of an understanding of mathematical structure involves determining what is preserved and what is lost between specific structures which have some isomorphism (e.g., Getner & Markman, 1994; Halford 1993). An example of such a structure is the commonality between the multiplicative structure of whole numbers, decimal numbers, common fractions and ratio, commonly called the mapping instruction approach (Peled & Segalis, 2005).

A variety of representational systems (e.g., iconic, verbal, movement, graphical and symbolic) augment mathematical understanding. We conjecture that allowing students to compare and move through a variety of representational systems supports the emergence of algebraic thinking. No one representational system provides all of the answers. Each has its limitations and strengths. Thus as we engaged in classroom conversations with young students we were continually exploring new signs that would assist students to extract the essence of the mathematics embedded in the exploration. Radford (2005) referred to this as a semiotic node, the point where all signs interact to reveal the mathematical essence of the concept. Instead of viewing representations progressing from the enactive, to iconic to symbolic as the individual translates experience into the model of the world (Bruner, 1966), we viewed mathematical development as cumulative, rather than as a replacement of earlier ways of thinking. Viewing cognitive development as cumulative allows one to flexibly traverse between 'levels' of representation as one encounters obstacles requiring a reconsideration and reconstruction of earlier ideas. Expression and language are essential to this journey as they give subtle shades of meaning that arise from human thought (Tall, 2004). From our perspective representations are not taken as a given. They assist us to arrive at some mathematical certainty about the situation we are investigating. As Smith (2006) states the representation becomes part of the knowledge of the learner; it is an integral component of the objectification process.

THE CASE OF EQUIVALENCE

With regard to equivalence, past research has evidenced that students posses a narrow and restricted knowledge of arithmetic at an early age (e.g., Carpenter, Franke, & Levi, 2003). An example of this is young children's understanding of the equal sign. Many interpret it as meaning "here comes the answer" (Carpenter et al 2003; Warren, 2002). A further example is young children's limited understanding of "turn arounds". When 73 Year 3 students were asked if $2 + 3 = 3 + 2$ and $2 - 3 = 3 - 2$ were both true, only 25%

indicated the first equation was true and the second false (Warren, 2002). The results from this study indicated that teaching materials themselves can act as cognitive obstacles to abstracting the underlying structure of arithmetic. Many of these students had very limited experience with equations written in the horizontal format. The position of the equal sign also caused difficulties. This was supported by past research which found that many children in elementary grades generally think that the equal sign means that they should carry out the calculation that precedes it and the number following the equation sign is the answer to that calculation (Saenz-Ludlow & Walgamuth, 1998).

Many young children also interpret "equals" as an indicator to do something (Behr, Erlwanger & Nicols, 1980; Carptenter & Levi, 2000). With regard to quantitative sameness, "equals" means that both sides of an equation are the same and that information can be gained from either direction in a symmetrical fashion (Kieran & Chalouh, 1992). Most students do not have this understanding; rather they have a persistent idea that the equals sign is either a syntactic indicator (i.e., a symbol indicating where the answer should be written) or an operator sign (i.e., a stimulus to action or "to do something" (Saenz-Ludlow & Walgamuth, 1998; Behr et al 1980; Warren, 2001). This incorrect understanding of "equals" appears to continue into secondary and tertiary education (Baroody & Ginsburg, 1983; Steinberg, Sleeman & Ktoriza, 1990) and seems to affect mathematics learning at these levels.

The model/representation/metaphor chosen for the equivalence and equation component of EATP was the balance model. Physical balance scales represented the notion of balance, and weights represented numbers. The theory that drove this decision was that of Davydov (1975). Generalising in numberless prior to numbered contexts enables investigation of concepts and structures not traditionally discussed until students have extensive knowledge of number (Davydov, 1975). We used Davydov's theory to assist us in establishing the structure of equations with equals as balance, equivalent class properties (symmetry, reflexivity, transitivity), and equations as complex structures (e.g., $5 + 2 = 6 + 1$). While past research has indicated that the balance model has its limitations (Aczel, 1998), it also has its advantages in that it considers both the right hand and left hand sides of equations and is not directional in any way (Pirie & Martin, 1997). Its limitation lies in its inability to model subtraction equations or unknowns as negative quantities, and thus is not seen as fostering a frame of mind adequate for the full development of algebraic thinking (Aczel, 1998). However, concrete models are endowed with two fundamental components (Filloy & Sutherland, 1996), namely, translation and abstraction.

Translation encompasses moving from the state of things at a concrete level to the state of things at a more abstract level. The model acts as an analogue for the more abstract. Abstraction is believed to begin with exploration and use of processes or operations performed on lower-level mathematical constructs (English & Sharry, 1996; Sfard, 1991). Through expressing the nature of their experience and articulating their defining properties, learners can construct more cohesive sets of operations that subsequently become expressions of generality. It is conjectured that abstraction entails recognising the important relational correspondence between the balance model, the weights as numbers, stating the identity of the two quantities or expressions (i.e., the two sides of the balance scales), and applying this understanding to more abstract situations, such as, equations involving subtraction and unknowns as negative numbers. Filloy and Sutherland (1996) suggest that not only do models often hide what is meant to be taught, but also present problems when abstraction from the model is left to the student. Teacher intervention is believed to be a necessity if the development of detachment from the model to construction of a new abstract notion is to ensue.

The language utilised during the teaching phase reflected the balance model with equal as balanced and having the same value on each side of the balance scale and unequal as unbalanced and having different values on each side of the balance scale. The symbols used were '=', '≠', and equations represented horizontally, with more than one value following the equal sign. The next section presents a journey of learning by the students as they progressed from Year 2 to Year 6. Particular emphasis was placed on the cognitive obstacles that occurred during this time and classroom actions, signs and representations that began to assist these students to continue their journey.

DESIGN OF EATP

The methodology adopted for EATP was longitudinal and mixed method using a design research approach, namely teaching experiments that followed a cohort of students over a four year period (Year 2 to Year 6). In line with this approach, during and between lessons hypothesis were conceived 'on the fly' (Steffe & Thompson, 2000) and modifications in the design were responsive to observed actions and understandings of the teacher/researcher and the students. For example, many of the instructional tasks were generated prior to the teaching phase, but during lessons tasks were modified according to classroom discourse and interactions. New representations were introduced in order to challenge students' thinking and encourage them to justify their responses. EATP was based on a reconceptualisation of content and pedagogy for algebra in the elementary school.

It sought to identify the fundamental cognitive steps crucial for an understanding of equivalence and expressions.

The participants across the years were a cohort of students, and their teachers, from 5 inner city middle class Queensland schools. During the study the cohort of students progressed from Year 3 to Year 6, with a small pilot study in Year 2. In total 220–270 students and 40 teachers participated in the study. All schools were following the new Patterns and Algebra strand from the new Queensland Years 1–10 Mathematics Syllabus (Queensland Study Authority, 2004).

All lessons were taught by one of the researchers. Although teachers in Queensland are well credentialised (all had 4-year training, in line with Queensland policy), the mathematics' component of this training is small and, like most elementary teachers in Queensland, they were not confident in teaching mathematics (Nisbet & Warren, 2000). In addition to this, algebraic thinking is a new content area in the elementary classroom requiring thinking that has not previously been explored, and all participating teachers in the research were unsure as to how to conduct lessons focusing on this new content area.

Data was collected from multiple sources. These included videos of the classroom during the teaching phase. All lessons were videotaped using two cameras, one fixed on the teacher and class as a whole and the other moving around the classroom focusing on student's activity of interest. Data included interviews with a randomly selected group of students, field notes written during the teaching phase, pre and post tests, and all worksheets and materials completed by students.

In this paper, the results are chronologically organised, beginning with the pilot study with the year 2 students. As we progressed through the years we changed our discussions and representations according to the difficulties in understanding that occurred. Thus the next section not only presents an overview of students' development but also a map of various signs that assisted them on their journey and the interplay between learning and teaching.

RESULTS: THE CASE OF EQUIVALENCE AND EQUATIONS

Year 2 (average age 7 years)

In the initial stages of the research we were most interested in establishing the concept of equation as balance.

Equivalence Using Unmeasured Models

We began the discussion about equivalence with a collection of groceries of varying masses, objects with unmeasured quantities.

Students were asked to identify what was the same and what was different; a language building activity. Most could distinguish same from different (Warren & Cooper, 2003), and many used hefting to justify their responses (*They weigh the same, they are the same shape, they are the same colour*). Students were then asked to identify two objects that had the same (equal) mass and different (not equal) mass. They checked their guesses by placing the objects on either sides of a balance scale and verbally shared the *mass of the pasta plus the rice is the same as (equal to) the mass of the salt plus baked beans. Also the mass of the beans plus the flour is different from (not equal to) the mass of the sugar and the pasta.* While these equations are not correct in terms of the mathematical symbol system as it is a relationship between masses and not numbers, they did serve as an effective model for complex equations and informal discussions about the equivalent class properties. This thinking was then mapped onto number situations. The groceries were replaced by a blue and green set of glass seashells all having the same mass. Figure 1 presents the mappings that occurred.

The robustness of believing that the equal sign indicates where to place the answer is illustrated by the following two excerpts from classroom conversations.

First, when Adam created the equation 2 + 2 = 2 + 2 on the balance scales:

Adam: Two and two equals two and two.

Martin: That is eight because four and four equals 8 [picking up all the materials and grouping them together.

Martin's actions and explanation reflected a deep rooted need to 'join' materials in addition situations, a common context to which many young

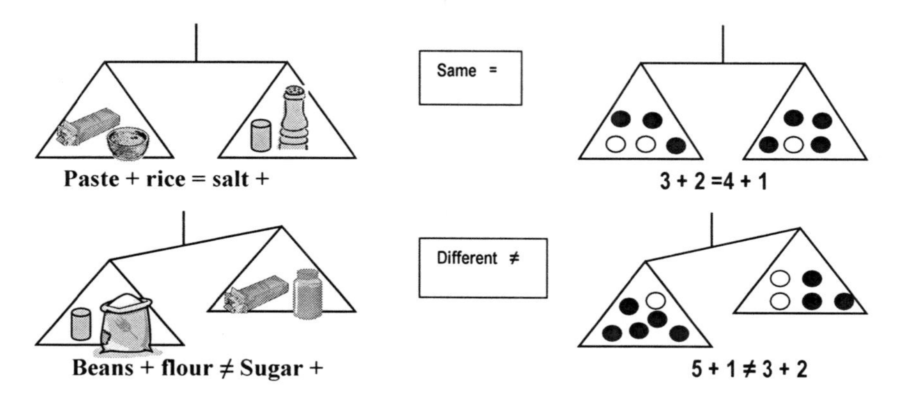

FIGURE 1. Mapping real world relationships to number situations.

students are exposed at an early age. We conjectured that this difficulty could be more than seeing 'equals' as a place to put the answer. It could include a perception of the addition sign as a signifier to 'join'. In our current context, most early year addition experiences involve modelling situations requiring joining sets of objects and answering questions such as 'how many birds are there altogether". We believed that this focus reinforces addition as a closed set requiring one to ascertain the total number of objects once the sets have been joined. While this is helpful for arithmetic thinking it is not helpful for early algebraic thinking. I return to this conjecture later in this paper.

Second, when the teacher/researcher wrote $3 + 4 = 6 + 1$ on the board and asked if this is true or not true:

> Bill: They aren't the same number because 3 and 3 equals 6: 3 plus 4 does not equal 6.
> T/R: Does everyone agree?
> John: No. $3 + 4 = 6 + 1$ because 3 plus 4 equals 7 and 6 plus 1 equals 7. They are the same because $3 + 4$ is the same as $6 + 1$.

From this excerpt it appears that the use of a balance metaphor and the development of mathematical language gave students a platform to challenge thinking, especially thinking that viewed equals as a syntactic indicator for a place to put the answer. It assisted in revealing the structure of equations.

Year 3 Classroom (average age 8 years)

The main focus of EATP with these students was to introduce the unknown into a discussion about addition situations and to investigate using the balance principle to solve for the unknown. All classroom discussions were supported by concrete materials. Due to our limited access to a large collection of glass beads we chose two sets of small cans all weighing 125 g as our concrete materials.

The Balance Principle and the Unknown

Before introducing unknowns, the discussion centred on "keeping the scales balanced" and the generality; if you add or subtract any number from one side of the balance scale you need to add or subtract the same number from the other side to keep the scales balanced - the balance principle. The unknown was introduced as an empty cloth bag with questions marks on it. The teacher secretly placed the value of the unknown in the bag as different equations were modelled. All students found it easy to solve equations

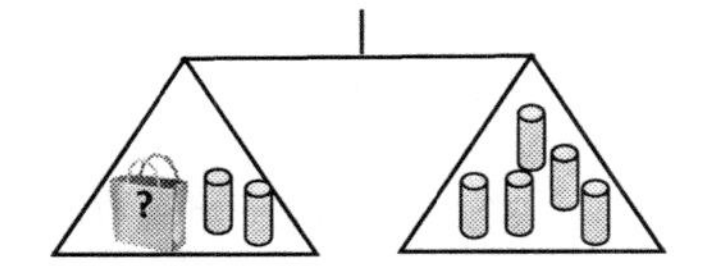

Oscar: If you take the bag from one side then you keep taking cans from the other until it is balanced. That is the unknown

Jill: Take 2 cans from both sides so the unknown is 3.

FIGURE 2. Model for ? + 2 = 5 and two typical solution paths

such as ? + 2 = 5. Figure 2 illustrates the model used in the classroom and the typical discussion that ensured.

Initially many students shared that there *are three things in the bag because three and two make five,* a reliance on number facts. While this is helpful for solving problems in an early elementary context, this thinking is based on carrying out a computation, arithmetic thinking. From our view, early algebraic thinking is about process, a focus on principles used to ascertain the answer. Thus students were encouraged to justify their responses by using the balance principle, thinking that was not reliant on number facts. Algebra is not about guessing but using signs in distinctive ways.

At the completion of the teaching phase 13 students were randomly selected for a one-on-one interview. The purpose of this interview was to gauge their understanding of the balance principle in addition situations. The interview included the equation ? + ?+ 2 = ? + 5, an equation that did not form part of the teaching seqence. This equation represents an example of what Filloy and Rojano (1989) refer to as the didactic cut between arithmetic and algebra, a first-degree equation with unknowns on both sides. They suggested additional resources are needed to overcome this barrier. The purpose of its inclusion in the interview was to see if the use of this representation (the balance model with the balance principle) assisted them in the beginning to cross this barrier. Our primary question was: Could these students spontaneously use their 'primitive' understanding of the balance principle to solve more complex problems?

Three students successfully obtained the correct answer. These students all used the balance method to find the unknown.' The following extracts from the transcripts delineate how these students reached solutions for this problem:

Used Balance to Find the Unknown

Elizabeth: That's hard; I will use the balance scales. There must be one baked bean in the mystery bag and one baked bean in the other bag and that equals 2, No, plus 2 so that has to be.... we don't know what these are just yet. If I took away 2 from 5 that would give me 3 [?+?=?+3] so there is 3 in one of these mystery bags. I think there is 3 in each mystery bag.

Sue: This is hard. Can I move the side around? I could do 5 plus 2 equals 7. So 7 plus 2 plus 5 and then found out what that answer is and that could be the one of these. So that could be 2 plus 2 plus 2.

Interviewer: Yes but we have to get it to balance. There is a way to do it to find out what they are.

Sue: Take away 2. And put a take away 2 at the end. So 5 take away 2 equals 3 so one of those question marks might be 3. 5 take away 2 equals 3... 5 plus 3 equals 8 so 3 plus 3 plus 2 equals 8.

In both cases these students only used the balance to remove the 2, giving ? + ? = ? + 3. From this pattern they seemed to simply see that each '?' must be 3 for the equation to be correct.

At the end of Year 3 an analysis of the classroom transcripts indicated that there were still a number of students who had the propensity to 'close' on the equations, to find an answer by adding all the components. As indicated earlier we conjectured that this was closely related to the focus in our elementary classroom on addition as joining. Thus the focus in the Year 4 classroom was to create addition and subtraction problems that involved comparing two situations to ascertain if they were equivalent.

Year 4 (average age 9 years)

Equivalent Stories

Given our concerns about the students' continual propensity to find answers to addition problems and our conjecture that this resulted from a view of addition as joining, in a review lesson students were asked to create equivalent stories involving addition. Figure 3 illustrates some typical stories.

The creation of stories did not involve finding answers but rather focussed on whether the contexts were equivalent appeared to assist these students move beyond the continual need to close addition and subtraction situations. We suggest this lesson assisted students to construct a new mental model that incorporated addition and subtraction situations that were equivalent.

Moving to Subtraction

The literature suggests that the balance model's limitation is its inability to be utilised in subtraction situations. We endeavoured to overcome some of these limitations by developing a representation that maintained some of the balance model features. The new representation was constructed from magnetic strips which allowed students to move the balance up and down as they added or subtracted numbers from each side. This maintained the

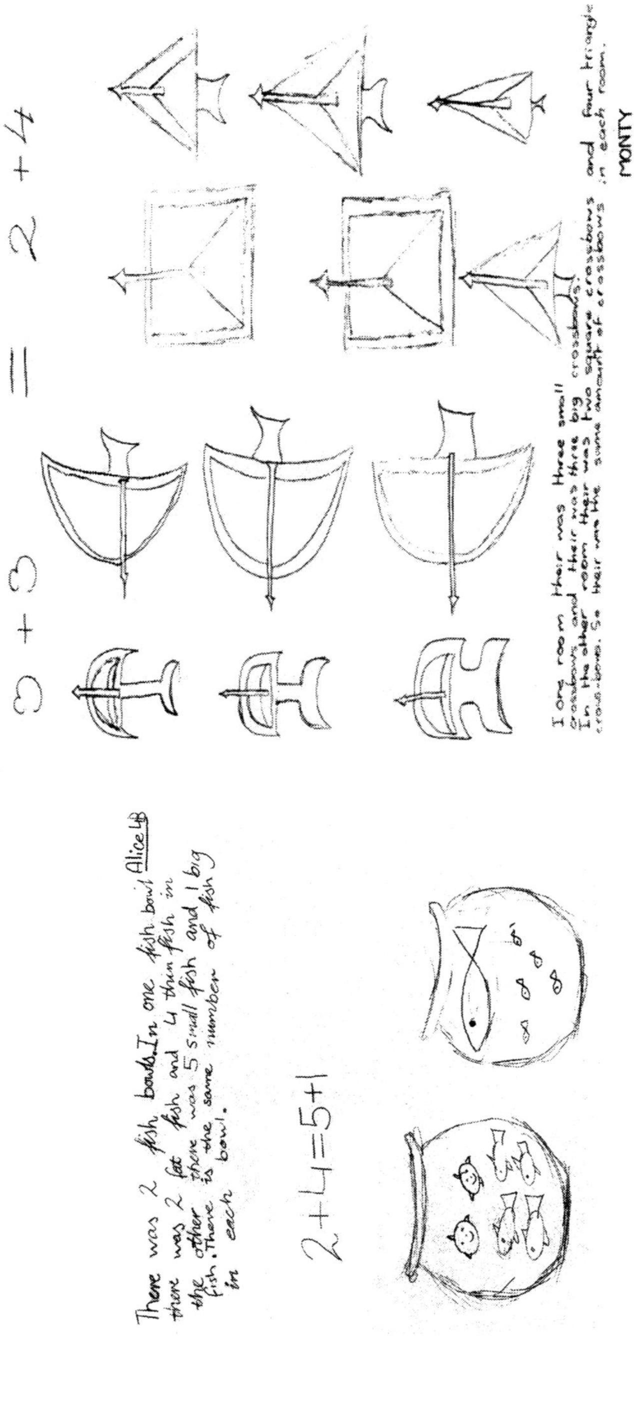

FIGURE 3. Equivalent stories.

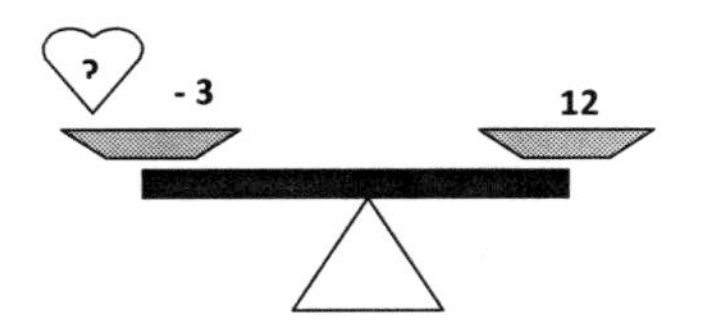

Mat: You could take three from both sides.

T/R: How many would you have on this side (pointing to the LHS)

Mat: Unknown.

T/R: How many on this side (pointing to the RHS)

Mat: 9.

FIGURE 4. Materials used for subtraction situations and a typical solution.

metaphor of movement up and down indicating balance or imbalance (equal or not equal). The unknowns were represented as magnetic shapes with question marks.

As we moved into subtraction and mapped the balance model onto our dynamic representation, many students experienced difficulties. Figure 4 illustrates the model we utilised for subtraction situations and the difficulties that students experienced with subtraction contexts.

The initial difficulties did not appear to emulate from the model itself but rather from these students' limited understanding of addition and subtraction. It became obvious in our discussions that many did not explicitly know that addition and subtraction were inverse operations or that if you subtract and then add the same number to another number then the number remains unchanged, the identity for addition and subtraction. We initially 'acted this out' using concrete materials. Figure 5 presents the sequence used in this discussion.

This discussion continued to "If I add any number of counters to your counters and then subtract the same amount, how many counters would you have?" Modelling subtraction proved more difficult. The students insisted that you could not subtract a number that was larger than their initial number of counters.

The Expression/Equation Dilemma

The generalisation of the balance principle is that equations remain the same if you 'do' the same thing to both sides. Thus inherent in the balance metaphor is the notion of sameness. In contrast the inverse relationship

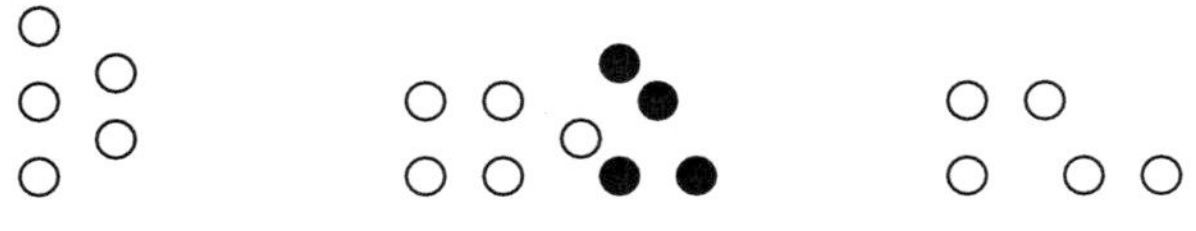

If you have 5 counters and I give you 4 more and I take 4 back how many do you have?

FIGURE 5. Acting out the additive identity.

between operations requires a metaphor of opposites. For example, for the equation ? + 8 = 12, isolating ? requires the inverse operation (subtracting 8) while the balance principle requires the same operation (subtracting 8 from both sides). These conflicting metaphors resulted in confusion for many of these students (we call this a compound difficulty). This difficulty became more evident as we changed to our new model.

To assist in addressing this compound difficulty we introduced the students to the number line. This representation allows one to model addition and subtraction as moving backwards and forwards on the line. Some initial questions were, "If you are at 8 on the number line and add 3, where will you be? How do you get back to 8?"The number line we used for discussions about unknowns was an open number line. Students were given an open number line with an 'unknown' marked on it and asked: "If you were at an unknown place on the number line and you added 3 to where you were, where would you be now? How would you get back to where you were?" Figure 6 present two typical responses to these questions.

Initially some students experienced difficulty with this. It was not until we drew an unmarked number on the floor and they physically stood on the 'unknown' spot, walked three paces and then walked back three paces did they begin to realise that 'addition is the reverse/inverse of subtraction' and 'subtraction is the reverse/inverse of addition.' Figure 6 presents some typical responses from their worksheets.

We had underestimated the importance of kinaesthetic movement and gestures to the development of mental models. As Radford (2006) succinctly claims, the perceptual act of noticing unfolds in a process mediated by multi-semiotic activity (e.g., spoken words, gestures, drawings etc.). From our experience, kinaesthetic movement is also an important inclusion to this list. In fact, at the completion of the lessons many students engaged in a process of walking up and down the number line until they were satisfied that they 'understood' the relationship between addition and subtraction and the notion of the identity. While the idea of adding zero to a number leaves the number unchanged appears trivial, the notion of adding something to a number and then subtracting that same amount from the answer results in the original number was not.

Co-variational Relationship between Two Unknowns.

One of the difficulties that many adolescent students experience is the relationship between two variables in the one equation (Fuji, 2003). Fuji found that many university students believed the same variables in an equation can simultaneously be different values and that different variables in an equation can never be the same value. To begin to address this issue we introduced discussion about equations with two unknowns and the relationships that existed between these unknowns. Equations were modelled

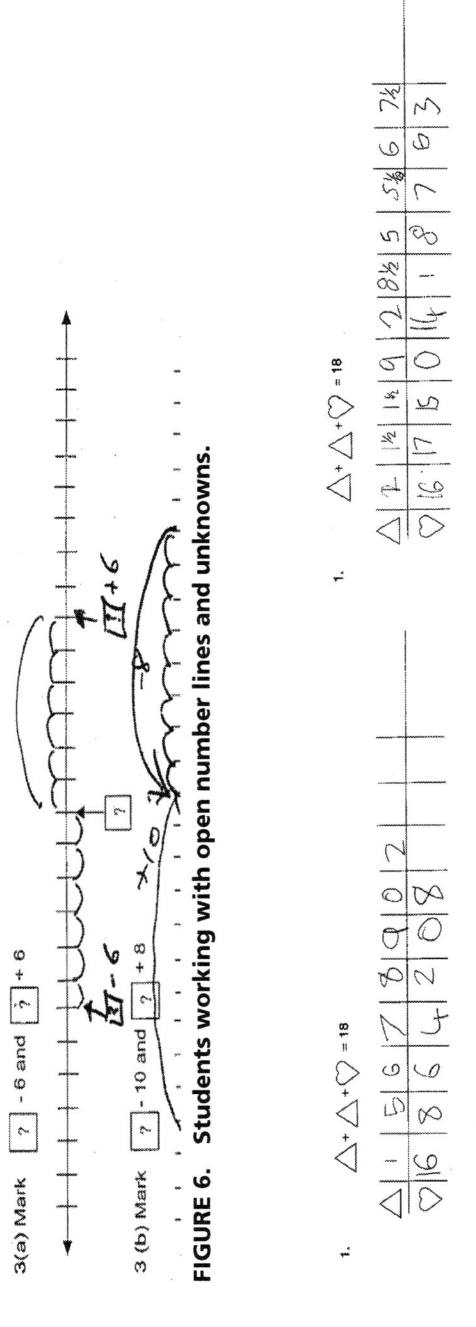

FIGURE 6. Students working with open number lines and unknowns.

FIGURE 7. Responses to the question ▲ + ▲ + ♥ = 18.

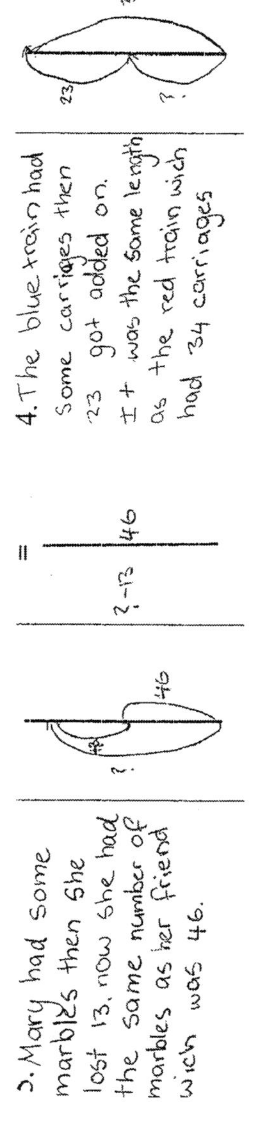

FIGURE 8. Some typical stories for addition and subtraction.

using different shaped boxes to represent the unknowns and marbles to represent numbers. Initially students were given equations such as ♥ + ■ = 8 and asked to find all the different values for each unknown and record their answers in a table of values. This thinking was extended to larger numbers. The discussions focussed on the different values each can hold, the processes use to record these values, the introduction of a table of values into the conversation. On an accompanying work sheet they were asked to find the values of ▲ and ♥ for the equation ▲ + ▲+ ♥ = 18. Figure 7 presents two student's responses.

The above responses show that not only can young students begin to explore the co-variational relationships between two unknowns but also extend this thinking to situations involving rational numbers.

In order to re-examine the compound difficulty students experienced with the balance principle and inverse principle in Year 5 we decided to introduce a new model for exploring equations, the length model. The advantage of the length model is that it allows modelling of both addition and subtraction situations. Its disadvantage lies in representing unknowns as unknown length and representing numbers as known lengths simultaneously, a visual that incorrectly encourages students to find the length of a unit from the known and apply this thinking to solve the unknown.

Year 5 (average age 10 years)

The Length Model for Discussion about Addition and Subtraction.

In the first stage, students were encouraged to write stories involving unknowns, model these stories with strips of papers and finally write these stories as equations. Figure 8 illustrates some typical addition and subtraction stories.

Most students could successfully model addition and subtraction problems using the length model. We conjecture that their previous work with the balance model and the number line assisted this transition. The length model appeared to be a vehicle that enabled students to easily amalgamate the two principles, isolating the unknown and maintaining 'balance' or

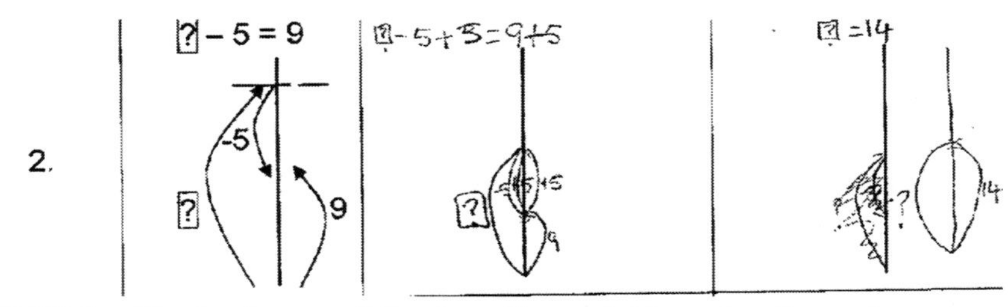

FIGURE 9. Students use of the length model to solve for unknowns.

sameness. Figure 8 demonstrates their ability to transfer this understanding to solving equations with unknowns.

The length model proved useful as an iconic representation for the 'melding' of the two conflicting principles, suggesting that integrating models results in a greater understanding of concepts.

Year 6 classroom (average age 11 years and 6 months)

Solving Equations with Unknowns on Both Sides

In the final year of the project we moved towards using the balance principle to solve complex equations with more than one unknown on each side. Our particular questions were (a) did the use of the physical balance model give students an appropriate mental model that allowed them to bridge the cognitive gap/didactic cut, and (b) did their mental model of the balance principle assist students in 'seeing' arithmetic in a way that allowed them not only to solve algebraic equations but also 'justify' their actions.

On the board I wrote $5 \times ? + 22 = 7 \times ? - 2$ and called upon volunteers to help me model this equation and find the solution for the unknown. The following classroom discussion occurred.

T/R: How do you model this with the balance scale?
Gemma: 5 times ? is 5 unknowns so you need 5 of them [whispering - multiplication is repeated addition].
John: And you need 7 of them [pointing to the 7 × ?]

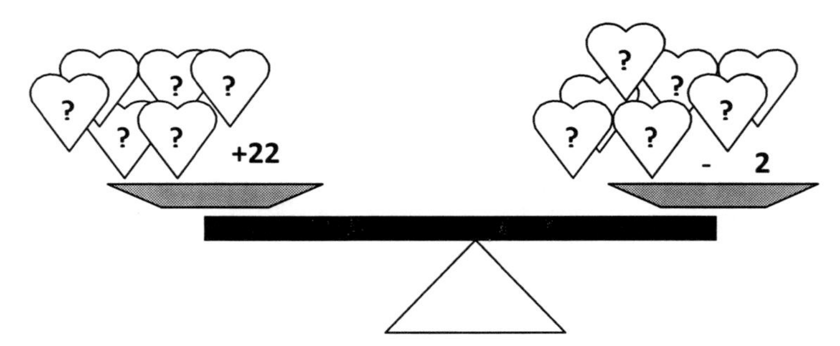

FIGURE 10.

On the whiteboard they used the magnetic materials to create the following representation of the equation.

T/R: What can you do to both sides and leave it balanced?
Adam: Add 2

T/R: Can you see anything that you can take from both sides?
Gemma: Four unknowns [taking 4 unknowns from both sides]
T/R: What else could you do to both sides?
Gemma: Take another unknown.
T/R: What does the equation now say?
Gemma: 22 equals 2 times the unknown minus 2.
T/R: How can you get the unknown by itself?
Gemma: Plus 2 to both sides [whispering plus 2 is the opposite of minus 2]
Gemma: 24. 24 is equals 2 times the unknown.
T/R: How do you solve for the unknown?
Gemma: You divide by 2—the unknown is 12.

The initial difficulty that these students experienced was extending the balance principle to unknowns, that is, taking the same unknown from each side left the equation balanced. But as evidenced by Gemma's response these students were capable of using the balance model to successfully solve equations with unknowns on both sides.

Solving Simultaneous Equations

The final challenge was to present two simultaneous equations and ascertain if these students were capable of using their understanding of the principles and properties of arithmetic that they had developed over the last 4 years in new unseen contexts. The problem on the work sheet was:

The store sells circles and triangular boxes. Write numbers in the shapes that make the equations true.

$$\blacktriangle + \bullet = 47$$

$$\blacktriangle - \bullet = 7$$

Most students could engage with this problem. Some used guess and check while twenty percent of the students could utilised algebraic means. They could also justify their thinking. Figure 11 presents two solutions proffered by these students.

These students still have one year of elementary school to complete before they begin to engage in a formal algebra course.

In summary, our results in the equivalence and expression dimension of EATP are indicating that very young students can:

1. represent equivalence in equation form in un-numbered and numbered situations;
2. generalise the equivalent class principles for equivalence; and
3. generalise the balance principle for simple equations.

It is also showing that older students can:

1. represent equivalence with unknowns in equations form;
2. generalise the balance principle for all operations; and
3. use the balance principle to solve unknowns in linear equations, including equations with unknowns on both sides.

CONCLUSIONS AND IMPLICATIONS

The conclusions from this research are presented under three main themes, namely, elementary students' ability to engage with algebraic thinking, the representations and teacher actions that support this thinking, and young students' development of mathematical structure.

First, EATP demonstrates that students can learn to understand powerful mathematical structures, usually reserved for secondary school, in the early and middle years of elementary school if instruction is appropriate. They can engage in conversations and present generalisations and justifications that evidence that they are capable of working effectively with unknowns that represent one value, unknowns as variables, situations with more than one unknown, and even start to engage in discussions about solving simultaneous equations. However, in all instances the contexts were firmly grounded in real-world situations and symbolic manipulation was kept to a minimum. EATP certainly assisted these students to acquire a broader understanding of equals and arithmetic in general. The results also evidence that the intertwining of arithmetic thinking and algebraic thinking (as defined by Malara and Navarra) certainly had "payoffs" for both (Mason, 2006). As students developed a deeper understanding of the structure of arithmetic they began to search for generalisations in computational contexts. In particular they could engage with the principles associated with the equivalence class and field structures.

Second, the position that learning is connections between representations (Radford, 2006; Hiebert & Carpenter, 1992), conversions using different representations, and flexible movement between representations at opportune times certainly enhanced student learning. Central to this is the socio-constructivist theory of learning, inquiry based discourse and the simultaneous use of multi-representations to build new knowledge (Warren, 2006). The major representations used in EATP worked, particularly in the way sequences of representations were used from acting out with materials through diagrams to language and symbols. But fundamental to this process was the role of the teacher. As Filloy and Sutherland (1996) claim, teacher intervention is necessary to assist in the appropriate detachment and abstraction from the model to objectifying the mathematics inherent in the representation. Learning can also be enhanced by creative representation-worksheet partnerships. Often teachers restrict worksheets to the symbolic register. EATP has evidenced that creative use of pictures

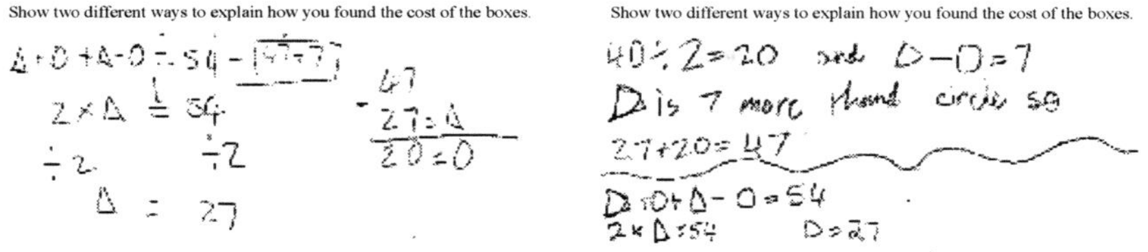

Show two different ways to explain how you found the cost of the boxes.

Show two different ways to explain how you found the cost of the boxes.

FIGURE 11. Student's solutions for solving the two equations.

and directions can allow a worksheet to reinforce understandings (as well as procedures) and to highlight principles.

Third, a teaching focus on structure is a highly effective method for achieving immediate and long term mathematical goals. In particular, the following representation sets were very effective in motivating students, solving problems and building principles and structure: (1) beam balances, cloth bags and objects and their pictures, and (2) walking games, paper strips and number lines. Some activities necessary for building structure also affect cognitive load. This is particularly so when large numbers are used to prevent guessing and checking as a strategy for determining answers and to direct students towards the principle.

In EATP, because of the separate focus on relationship through equations and change through function machines, there was overlap with regard to the identity and inverse principles that reinforced these structures in both perspectives. At the beginning of our journey we taught equivalence and equations in the first half of the year and functions and patterns in the second half. This artificial separation of these two areas and the sequence in which we chose to engage these ideas caused us difficulties. However as we reached the modeling of subtraction situations there was a need to introduce thinking of arithmetic as change and reversing change. The compound difficulty or mixed metaphors of sameness and opposites required us to introduce 'change' to help us understand relationships. One consequence was that in subsequent years we introduced functional thinking before we began discussions about equivalence. This example evidences the compounding effect of building structure through small steps. It is necessary to build a superstructure into which to place conflicting principles such as compensation for addition and subtraction and inverse and balance for solutions of linear equations.

Finally, the conversations presented in this paper begin to tease out some of the key issues and learning that need to occur in the elementary years from an early algebraic thinking perspective. For the equivalence and equation strand these are, an understanding of equations as representing equivalent situations, the ability to apply the balance principle to solve for unknowns, an understanding of the sign systems we use to represent unknowns (and variables), recognizing the identity and inverses for all operations, and finally the ability to use this thinking to find solutions for and generalizations about real world problems involving more than one unknown. The conversations also delineate representations and teacher actions that begin to support student learning. As mentioned earlier, this research is exploratory by nature and as such commences to map the territory. It begins to sketch out a learning-teaching trajectory for early algebra. Much needs to be done to fill in the gaps.

I will conclude with a quote from Mason (2006), a value that underpins all our work in this area.

> No one should be confined to arithmetic calculations as their basic numeracy because it stunts or even blocks their access to the kind of thinking that is essential for participation in a democratic society (Mason, 2006, p. 79).

REFERENCES

Aczel, J. (1998). Learning algebraic strategies using a computerised balance model. In A. Oliver & K. Newstead (Eds.), *Proceedings of the 22nd conference of the International Group for the Psychology of Mathematics Education* (Vol. 2, pp. 1–16). South Africa: Stellenbosch.

Baroody, A. J., & Ginsburg, H. P. (1983). The effects of instruction on children's understanding of the equals sign. *Elementary School Journal, 84,* 199–212.

Behr, M., Erlwanger, S., & Nichols, E. (1980). How children view the equals sign. *Mathematics Teaching, 92,* 13–18.

Bruner J. S., (1966). *Towards a Theory of Instruction.* New York: Norton.

Carpenter, T., Franke, M., & Levi, L. (2003). *Thinking mathematically: Integrating arithmetic and algebra in elementary school.* Portsmouth: Heinemann.

Clements, D., (2007). Curriculum research: Towards a framework for "Research based curricula." *Journal for Research in Mathematics Education, 30,* 35–70.

Davydov, V. V. (1975). The psychological characteristics of the "prenumeral" period of mathematics instruction. In L. P. Steffe (Ed.), *Children's capacity for learning mathematics* (Soviet studies in the psychology of learning and teaching mathematics, Vol VII, pp. 109–205). Chicago: University of Chicago.

Dienes Z. P. (1961). *On abstraction and generalisation.* Boston: Harvard University Press.

English, L, & Halford, G. (1995). *Mathematics education: models and processes.* Mahwah, New Jersey: Lawrence Erlbaum.

English, L. D., & Sharry, P. (1996). Analogical reasoning and the development of algebraic abstraction. *Education Studies in Mathematics, 30,* 135–157.

Filloy, E. and Rojano, T. (1989). Solving equations: The transition from arithmetic to algebra. *For the Learning of Mathematics,* 9(2), 19–25.

Filloy, E., & Sutherland, R. (1996). Designing curricula for teaching and learning algebra. In A. Bishop, K. Clements, C. Keitel, J. Kilpatrick & C. Laborde (Eds.), *International Handbook of Mathematics Education* (Vol. 1, pp. 139–160). Kluwer: Dordrecht.

Fujii, T. (2003). Probing students' understanding of variables through cognitive conflict: is the concept of a variable so difficult for students to understand. In N. Pateman, B. Dougherty, & J. Zilliox (Eds.), *Proceedings of the 27th conference of the*

International Group for the Psychology of Mathematics Education, (vol. 1, pp. 47–65). College of Education: University of Hawaii.

Getner, D., & Markman, A. B. (1994). Structural alignment in comparison: No difference without similarity. *Psychological Science, 5*(3), 152–158.

Halford, G. S. (1993). *Children's understanding: The development of mental models.* Mahwah, NJ: Lawrence Erlbaum.

Hiebert, J., & Carpenter T. P. (1992). Learning and teaching with understanding. In D. Grouws (Ed.), *Handbook for research on mathematics teaching and learning.* (pp. 65–97). New York: MacMillan.

Kaput, J. (2006). What is algebra? What is algebraic reasoning? In J.Kaput, D. Carraher, M. Blanton (Eds.), *Algebra in the early grades.* (pp. 5–17). Mahwah, NJ: Lawrence Erlbaum.

Kieran, K., & Chalouh, L. (1992). Prealgebra: The transition from arithmetic to algebra. In T. D. Owens (Ed.), *Research ideas for the classroom: Middle grades mathematics* (pp. 179–198). New York: MacMillan.

Linchevski, L. (1995). Algebra with number and arithmetic with letters: A definition of pre-algebra. *Journal of Mathematical Behaviour, 14,* 69–85.

Linchevski, L., & Herscovics, N. (1996). Crossing the cognitive gap between arithmetic and algebra: Operating on equations in the context of equations. *Educational Studies, 30,* 36–65.

Malara, N., & Navarra, G. (2003). *ArAl Project: Arithmetic pathways towards favouring pre-algebraic thinking.* Bologna, Italy: Pitagora Editrice.

Mason, J. (2006). Making use of children's powers to produce algebraic thinking. What is algebra? What is algebraic reasoning? In J.Kaput, D. Carraher, M. Blanton (Eds.), *Algebra in the early grades.* (pp. 57–94). Mahwah, NJ: Lawrence Erlbaum.

Nisbet, S., & Warren, E. (2000). Primary school teachers' beliefs relating to teaching and assessing mathematics and factors that influence these beliefs. *Mathematics Teacher Education and Development, 2,* 34–47.

Peled, I., & Segalis, B. (2005). It's not too late to conceputalise: Constructing a generalised subtraction scheme by abstracting and connecting procedures. *Mathematical Thinking and Learning, 7*(3), 207–230.

Pirie, S. E. B., & Martin, L. (1997). The equation, the whole equation, and nothing but the equation! One approach to the teaching of linear equations. *Educational Studies in Mathematics, 34*(2), 159–181.

Queensland Study Authroity (2004). *Mathematics years 1-10 syllabus.* Retrieved May 24, 2008, from www.qsa.qld.edu.au.

Raddford, L., Bardini, C., Sabena, C., Diallo, P., & Simbagoye, A. (2005). An embodiment, artifacts and signs: A semiotic-cultural perspective on mathematical thinking. In H. Chick & J. Vincent (Eds.) *Proceedings of the 29th conference of the International Group for the Psychology of Mathematics Education,* (Vol. 4, pp. 113–120). Melbourne: University of Melbourne.

Radford, L. (2006). *Algebraic thinking and generalization of patterns: A semiotic perspective.* Paper presented at the 28th annual meeting of the International Group for the Psychology of Mathematics Education (NA), Merida, Mexico.

Saenz-Ludlow, A., & Walgamuth, C. (1998). Third graders' interpretation of equality and the equal symbol. *Educational Studies in Mathematics, 35,* 153–187.

Scandura, J. M. (1971). *Mathematics: Concrete behavioural foundations.* New York: Harper & Row.

Sfard, A. (1991). On the dual nature of mathematical conceptions: Reflections on processes and objects as different sides of the same coin. *Educational Studies in Mathematics,* 22, *1–36.*

Skemp, R. (1978). Relational understanding and instrumental understanding. *Arithmetic Teacher, 26*(3), 9–15.

Smith, E. (2006). Representational thinking as a framework for introducing functions in the elementary curriculum. What is algebra? What is algebraic reasoning? In J. Kaput, D. Carraher, M. Blanton (Eds.), *Algebra in the early grades.* (pp. 133–160). Mahwah, NJ: Lawrence Erlbaum.

Steffe, L. P., & Thompson, P. W. (2000). Teaching experiment methodology: Underlying principles and essential elements. In A. E. Kelly & R. A. Lesh (Eds.), *Handbook of research design in mathematics and science education* (pp. 267–306). Mahwah, NJ: Lawrence Erlbaum.

Steinberg, R., Sleeman, D. & Ktorza, D. (1990). Algebra students' knowledge of equivalence of equations. *Journal for Research in Mathematics Education, 22*(2), 112–121.

Tall. D. (2004). Thinking through the three worlds of mathematics. In M. Hoines & A. Fuglestad. *Proceedings of the 28th conference of the International Group for the Psychology of Mathematics Education,* (Vol. 4, pp. 281–288). Bergen: Bergen University College.

Usiskin, Z. (1988). Conceptions of school algebra and uses of variables. In A. Coxford, & A. P. Schulte (Eds.), *Ideas of algebra, K–12* (pp. 8–19). Reston, VA: NCTM.

Van den Heuvel-Panhuizen. M. (2002). *Realistic Mathematics Education as work in progress.* Retrieved May 7, 2008 from http://www.fi.uu.nl/publicaties/literatuur/4966.pdf.

Warren, E. (2002a). Children's understanding of turnarounds: A foundation for algebra. *Australian Primary Mathematics Classroom, 7*(1), 9–13.

Warren, E. (2002b). Unknowns, arithmetic to algebra: Two exemplars. In A. Cockburn & E. Nardi (Eds.), *Proceedings of the 26th conference of the International Group for the Psychology of Mathematics Education.* (Vol. 3, pp. 369–376). Norwich: University of East Anglia, England.

Warren, E. (2003). Young children's understanding of equals: A longitudinal study. In N. Pateman, G. Dougherty, & J. Zilliox (Eds.), *Proceedings of the 27th Conference of the International Group for the Psychology of Mathematics Education* (Vol. 4, pp. 379–387). Honolulu, HI: College of Education, University of Hawaii.

Warren, E. (2006). Teacher actions that assist young students write generalizations in words and in symbols. In Novotná, J., Moraová, H., Krátká, M, & Stehliková, N. (Eds.), *Proceedings of the 30th Conference of the International Group for the Psychology of Mathematics Education,* (Vol. 5, pp. 377–384). Prague, Czech Republic: Charles University.

Warren, E., & Cooper, T. J. (2003). Arithmetic pathways towards algebraic thinking: Exploring arithmetic compensation in Year 3. *Australian Primary Mathematics Teacher, 8*(4), 10–16.

CHAPTER 3

A BRIEF ESSAY ON THE NEED TO CONSIDER THE "SUPERFICIAL" ASPECTS OF LEARNING ALGEBRA

Romulo Lins

INTRODUCTION

There should be no doubt that the mathematical education of a person has to aim at the development of a *deep* understanding of the subject (concepts, techniques, and applications). And this is true in particular for algebra education. Research and development in the field has been, and continues to be, concerned both with producing a sound understanding of teaching and learning processes and with producing the means through which that goal can be achieved.

However, the *superficial* side of the understanding of the subject has been left, I think, considerably unattended. By this I do not mean the mistakes people make when they transfer the *superficial* structure of, say, an algebraic transformation to another situation to which it does not apply. For instance, transferring:

Future Curricular Trends in School Algebra and Geometry: Proceedings of a Conference. pages 31–44

$$2(a+b) = 2a+2b$$

into

$$(a+b)^2 = a^2 + b^2.$$

With respect to that there has been plenty of documented research, particularly in the 1970s and 80s (see, for instance, Lins, 1992; Lins & Kaput, 2004).

What I do mean is that there are *superficial* aspects of learning mathematics that can and should be explicitly considered, but have not been. A good metaphor here seems to be this: if one dives into a lake, before reaching deep (and indeed even shallow) waters, it is necessary to go through the surface.

Let's consider a child who is learning to speak (a major, major, achievement). If she says, "Daddy, I maked a present for you!" it is highly unlikely that the father will offer any kind of correction for the misconjugated verb. Instead, it is very likely that he will be quite happy to notice that the child is developing a sense of time as related to language use. Is "maked" a sign of deep or of superficial understanding?

I am quite sure that all of us could come up with many similar examples.

Still, with respect to learning to speak a first language, most people do not treat it as if there was a correct or best sequence for mastering its different aspects; assessment of whether or not things are going well is made largely on the basis of how well the child functions socially. And, finally, hardly any person takes sole responsibility for a child learning to speak; brothers and sisters, grandparents and other relatives, neighbours, other children in various situations, TV, radio and so on, are all seen as part of this development. In other words, this is a massively collective enterprise (and thus *deeply* social and cultural).

School education, on the other hand, seems to take a different approach.[1] There are *better* sequences. Correcting *mistakes* is too often considered more important than encouraging new ideas. And even when one adopts a *spiral* curriculum, it is likely that a given teacher will indeed feel s/he is the only one responsible for getting the children adequately through a coil of the spring.

I do not intend to present even a simplified argument against such approach. What I do intend, though, is to take inspiration from how we relate with young children and their development outside school, and to propose that at least to some extent we can benefit from acting similarly in schools. After clarifying what this means I will offer a couple of examples of how it

---

[1] I am referring to *most* school systems. There are exceptions, but they are so rare as not to make much difference to my argument, unfortunately.

can be implemented in the classroom, arguing that what we propose is different from other, currently available, approaches, which may superficially look similar to it.

FRAMING THE ARGUMENT

Let's look again at the so well documented mistake:

$$(a+b)^2 = a^2 + b^2$$

Let's view it as the "maked" mentioned above. Doesn't it make sense to say that this could be an encouraging sign that the pupil is perceiving patterns, rather than a discouraging sign that the pupil is *not* understanding that a and b are numbers and that in general that equality does not apply? I think it does, it makes a lot of sense.

Maybe we are driven away from the former view because at the point pupils usually meet such statements we have already told them that letters, in the mathematics context, represent numbers, and so on, so we are *disappointed* that they do not take this into consideration and say.

$$(a+b)^2 = a^2 + b^2.$$

Take a minute to consider how to factor $x^3 + 1$. If you know the formula by heart (superficial), it will save you time. If you do not, it is possible that you say "hm, that *looks like* $x^3 - 1$ and this one I can handle", and ending up with $x^3 + 1 = x^3 - (-1)$ and so on. That *looks like* is precisely the kind of thing I am interested in, here.

On the other hand, the *mistake* mentioned above happens precisely because

$$2(a+b) = 2a + 2b$$

does *look like*

$$(a+b)^2 = a^2 + b^2.$$

Bring the two things together and you will get to my point: we need to educate our pupils' *perception* in algebra, the way they treat the *looks like* factor.

In this paper I will consider two aspects in relation to which this can be done quite early in school. First, the acceptance, as *legitimate* in school mathematics, of expressions involving numbers, letters and arithmetical operations (and eventually the equal sign). Second, the development of a notion of *form* in relation to algebraic expressions, including thinking of algebraic transformations in terms of change of *form*.

I chose these two because I think they are key parts of being fluent in algebraic manipulation. Again, I must remind the reader that I am *not* advocating mindless symbol pushing. I am simply claiming that unless those two aspects are contemplated it is hardly possible to use algebraic language transparently, and that is a big part of what I mean by *being fluent*.

HOW TO EDUCATE PERCEPTION?

A quote taken from Terry Wood, Megan Staples, Sean Larsen and Karen Marrongelle (2008):

> [...] One way to view the differences between mathematics and school mathematics is to describe them as disciplinary practices and learning practices following Cohen and Ball (2001). Consider for example justification and argumentation, these are disciplinary practices in mathematics, but in school mathematics these are learning practices. In mathematics justification and argumentation are disciplinary practices because they are the means by which mathematicians validate new mathematics. In school mathematics argumentation and justification are learning practices because they are the means by which students enhance their understanding of mathematics and their proficiency at doing mathematics.

I think a similar point can be made with respect to algebraic manipulation, but perhaps at a more basic level. In mathematics, algebraic manipulation is a tool that enhances one's ability to justify and argue. Algebraic manipulation happens, so to speak, in the background. In school mathematics, however, too often algebraic manipulation is in the foreground, that is, it is the very subject of study; it is not even a learning practice. How would it be possible to make it into a learning practice, into a *means by which students enhance their understanding of mathematics and their proficiency at doing mathematics*? Pushing it to the background seems a promising approach, as much as justification and argumentation happen, in school mathematics as well as in mathematics, largely in the background.

A second quote, taken from Anne Watson (2008):

> In this paper I argue that school mathematics is not, and perhaps never can be, a subset of the recognised discipline of mathematics, because it has different warrants for truth, different forms of reasoning, different core activities, different purposes, and necessarily truncates mathematical activity. ... The relationship of school mathematics to adult competence is similar to the relationship ... between being made to eat all your spinach and becoming a chef; between being forced to practise scales and becoming a pianist. There are some connections, but they are about having a focus on a narrow subset of semi-fluent expertise in negative social and emotional contexts, without full purpose, context and meaning. That some people become ... beautiful

pianists or inspiring cooks is interesting, but what is more interesting is the fact that most people who go through these early experiences do not: instead they merely follow orders, or hate green vegetables, or give up practising their instruments.

This is a quite interesting point: *school mathematics is not, and perhaps never can be, a subset of the recognised discipline of mathematics ... [it] necessarily truncates mathematical activity*. That is, maybe we cannot go too far in making school mathematics—or maybe better, school mathematical activity – look like mathematics or mathematical activity. But as I read about spinach and music a refreshing insight came to my mind. When we want children to learn to wear clothes, are we, to any extent, concerned with whether or not they will become clothing designers? When we want children to learn to eat using cutlery, are we, to any extent, concerned with whether or not they will become gourmets? I don't think so.

Could it be that if we serve spinach to a child early enough we wouldn't come to a point in which the child has to be *made to eat it*? That, instead, eating spinach would simply become something one does? And, again, can't we do so without being concerned at all about the child becoming a chef?

In Brazil, 7th grade (now renamed Year 8, with pupils around 14 years old) is when algebraic manipulation is treated *properly*, that is, becomes a subject of study. Before that, children have very little contact with literal notation in mathematics. To no one's surprise any longer, this is a grade in which the number of pupils failing is significantly higher than in the previous ones. The ever-repeating cry of despair from pupils is "Calculating with letters??". It seems they are being made to eat algebraic expressions and algebraic manipulation, and, of course, the reaction is similar to that of being made to eat spinach. But what if we served them algebraic expressions early, as with the spinach? Could it be that they would get used to them to the extent that they would become natural? And, notice, in doing so no one needs to be interested in whether or not our children are going to become mathematicians.

Yes, I am deliberately trying to be *superficial*.

From Alan Bishop (1994):

> Teacher training in mathematics involves much more than just learning how to manage a classroom effectively. Nor is it just a matter of learning sufficient mathematics to be able to teach that content to school students. Mathematics teachers are passing on values, habits and customs as well as knowledge and skills. They are inducting their students into the culture of mathematics. Culture is not being used here to refer to 'grand' culture, or 'high' culture (as in a 'cultured' person) but merely to reflect the fact that like language, religion or morals, mathematics is part of a culture's store of knowledge, developed by previous (and present) generations and made accessible to succeeding generations. ...

Before we teach children to eat using cutlery do we wait until they *properly* understand what microbes are? No. Before we teach children to use clothes do we wait until they properly understand what social conventions are? What laws are? Do we worry about allowing them to *construct* by themselves the concept of being clothed? No. Do they have to learn the importance of vitamins and iron to our bodies before we can serve them spinach? No, again.

Am I against active learning? No.

But I believe we can and should learn from the way people learn things outside schools. Not the deep aspects of such learning, but the superficial ones. For instance, it seems to be true that using cutlery and using clothes bring a better quality of *social* life for people. And it is for *this* reason that we teach them to our children before doing it makes much sense to them, and not because *potentially* they will be, one day, clothing designers or restaurant owners.[2]

Can we take a similar approach when considering algebraic expressions, for instance? I think we can. The difference might be that we take algebraic expressions into consideration because of something that will only happen in the yet not visible future, that is, learning and using algebra, while using cutlery and clothes and eating spinach are immediately visible as part of children's lives.

As to the title of this section (How to educate *perception*?) my first answer is this: as early as possible. But I would add that two points of view have to be taken into consideration, that of *normal* people and that of mathematics educators (teachers or not).[3]

What is it that *normal* people *see* when they see us *doing algebra*? Calculating with letters. Literally mixing letters with numbers. Mysterious (if not irrational) rules for doing it.

What is it that we, mathematics educators want them to *see*? Legitimate symbolic expressions. Meaningful transformations of those expressions.

I think that an answer to the question in the section title may come from considering both views at the same time: to educate their perception, in this case, means developing *legitimacy* for algebraic expressions and then developing *legitimacy* for expression transformation.

[2]The argument that we dress our children because of, say, the weather, is also interesting, but, again, this is a superficial (immediate). There are native Brazilians, for instance, who live in places where it gets quite cold in the winter, but instead of getting dressed they get used to it.

[3]By *normal* people I mean people who do not have any particular interest in mathematics (as it is the case with mathematics educators, mathematicians, engineers, and so on). *Normal* people have a functional interest in mathematics (everyday life). That means, almost surely, that *normal* people are not algebra users to any extent and, after leaving school (whatever the time they had spent there) gradually algebra is again, slowly or not, reclassified as something akin to ET language. *Normal* people include most children and teenagers.

In the next section I present a possible way to take a *naturalistic* approach in relation to algebraic expressions: if it succeeds, there will be a lot less "*being made to eat spinach*", which will possibly be replaced by "yummies" and "I prefer potatoes". Later in this paper, I extend the argument to the notion of *algebraic form.*

FRUIT SALAD CAN BE GOOD FOR ONE'S HEALTH!

My former PhD supervisor Alan Bell has many times said that representing "3 apples and 2 bananas" by "$3a + 2b$" is *fruit salad algebra,* and that is not a good thing because pupils will not learn that in school algebra letters stand for numbers, not things.

But what if I do not care that, at least at some point, they do not learn that in school algebra letters stand for numbers? Is there still something to be learned from doing fruit salad algebra? This section is an attempt to convince the reader that the answer is a quite important *yes.*

1. An Activity: A Weird Snacks and Soda Cans Shop

A shop sells packages of snacks and soda cans. You cannot buy separate snacks or cans in this weird shop... They sell the following packages:

Package A: 1 snack and 3 cans
Package B: 2 snacks and 3 cans
Package C: 2 snacks and 4 cans
Package D: 2 snacks and 6 cans
Package E: 3 snacks and 5 cans
Package F: 4 snacks and 3 cans

How can one buy...

3 snacks and 6 cans?
5 snacks and 8 cans?
1 snack and 1 can?
3 snacks and 9 cans?
(and so on)

To actually find the answers is not hard. It can be made harder with larger numbers or purchases that require multiple combinations. Children will be doing a lot of mental calculations (helpful). They will have to find ways of keeping track of combinations (quite helpful; they may want to make a table with all 36 2-packages combinations. What if I want to represent all 3-packages combinations? Maybe a table won't do, maybe it will...).

But our actual target, when we developed this activity, was something else:

> To buy 3 snacks and 6 cans one buys a package with 1 snack and 3 cans and a package with 2 snacks and 3 cans made into

3 snacks and 6 cans = (1 snack and 3 cans) + (2 snacks and 3 cans)

which easily turns into

$$3\ S \text{ and } 6\ C = (1\ S \text{ and } 3\ C) + (2\ S \text{ and } 3\ C)$$

and into

$$3\ S + 6\ C = (1\ S + 3\ C) + (2\ S + 3\ C)$$

Junk-food algebra? Well, not a nice name, but in Alan's sense, yes. What might we gain here?

At least two important things, I think. On one hand, pupils are operating with/on expressions *as whole objects*, supported by the *packages* context. On the other hand, literal expressions (such as $3\ S + 6\ C$) are *legitimate* as a notational aid. The abbreviation to S and C may come from the students (sometimes it does), but it may also come as a suggestion by the teacher (not to Vygotsky's opposition).[4] With second to fourth graders mixing letters with numbers with other mathematical signs and no sign of shock or despair, our objectives were reached.

As to the mathematical side of it, we find it hard to see it as anything else than polynomial addition. And in polynomials proper the *letters* are nothing but formal place-markers; (3, 6) instead of $3\ S + 6\ C$ would be equally fine. Not so bad: young children writing down and adding polynomials.

And they only had to do it this once to *learn* that doing so is *legitimate* in mathematics.[5]

2. *Another activity:* The Music Shop[6]

Daniel is helping his aunt, who owns a records shop (she went on a boat trip). When he gets to the shop, Monday morning, he re-

[4] We have never witnessed or got a report that a child complained of such abbreviation or said s/he had not *understood* it.

[5] Around 300 children in Brazil, grades 2–4, worked with this activity and were informally post-tested for the persistence of that *legitimacy*, using the activity presented on next section.

[6] Developed at the time when CD's and MP3 players were not available...

alises that he knew his aunt sold the records for a single price, and the tapes for a single price, too. But he forgot to ask what the prices were! Looking around he found a piece of paper with some of Friday's *sales*[7]:

1 record and 5 tapes — R$ 65
3 records and 4 tapes — R$ 85
2 records and 1 tape — R$ 40
4 records and 3 tapes — R$ 90
5 records and 2 tapes — R$ 95

Customers are arriving!! Let's help Daniel to calculate the cost of some new sales!

4 records and 9 tapes
4 records and 2 tapes
3 records and 1 tape
1 record and 1 tape

What is more expensive: a record or a tape?

Teachers suggested the following notation, which pupils readily accepted and used:

1 record and 5 tapes = R$ 65
+ <u>3 records and 4 tapes = R$ 85</u>
4 records and 9 tapes = R$ 150

quickly moving (on their own) to:

$$\begin{array}{rl} & 1\,R + 5\,T = \ \ 65 \\ + & \underline{3\,R + 4\,T = \ \ 85} \\ & 4\,R + 9\,T = 150 \end{array}$$

What's in it: (i) operating with and on expressions as whole objects; (ii) persistence of the legitimacy of expressions mixing numbers, letters, arithmetical operations and equality sign; and, (iii) more operations with/on expressions (+, –, x, ÷).

In this music algebra Alan's remark holds: the teacher has to make pupils aware that R stands for the price of a record and T for the price of a tape. Also, although it makes sense to write

$$3\,R + 2\,T = (6\,R + 4\,T) \div 2 = 130 \div 2 = 65$$

[7] Notice my didactical emphasis.

it does not make sense to write

$$(4\,R + 9\,T) \div 2 = 2\,R + 4.5\,T = 75$$

because 0.5 of a tape is not of much use. The *logic of the operations* is that of records and tapes, and the arithmetical calculations are only used to produce actual prices of sales. From our theoretical perspective that means there is no algebraic thinking going on (Lins, 1992, 2001).

The teacher may want to ask pupils if they can work out the price of a record and the price of a tape, but by design we always suggest that this is not done. The core of this activity, as in the previous one, is to deal with combinations of expressions.[8]

There are similar activities in which expressions like 4P – 3R are legitimate, and others in which, using decimal numbers, it is less easy to guess individual prices.

THE WATER TANKS

In the previous activities pupils could produce legitimate (for them) literal expressions and operate with/on them. The following activity adds the possibility of developing a sense of form and transformation of such expressions.

These are two identical water tanks. The tank on the left needs another 9 buckets full of water to fill it up. The tank on the right needs another 5 buckets full of water to fill it up. What can we say about this situation?

One can say that "the tank on the right has more water than the tank on the left, just look at the picture". Or that "the tank on the right has more water than the tank on the left, as less water is missing on the right one". Or that "the tank on the right has more water than the tank on the left, as only 5 buckets of water are missing in it, and 9 buckets on the right". From the perspective of our Model of Semantic Fields (Lins, 2001) these three statements *together* with the respective justifications consist in different *knowledge*. The *practical* consequence of including *justification* as a constitutive part of our definition of knowledge is that we can distinguish, in a sufficiently fine and simple way, different pieces of *knowledge* that involve the same statement (proposition).[9]

---

[8]We have suggested that teachers, in case pupils find out the individual prices, say that's fine but ask them if they can work out the cost of those sales without using them.

[9]We define *knowledge* as "a statement-belief (the statement of something in which a person believes) *together* with a justification that person has for believing in it." This definition immediately and clearly distinguishes the knowledge of a child and a mathematician who both say that " 3 plus 2 is equal to 2 plus three", the child's justification being showing it with her fingers and the mathematician's being that the addition of integers is commutative.

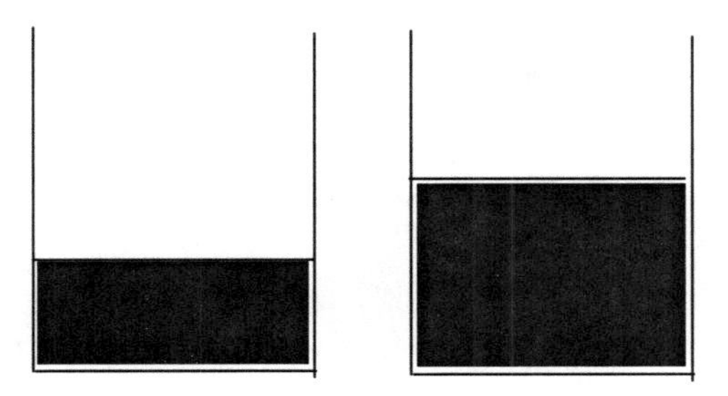

Water Tanks

In this activity we are concerned with the production of legitimate (for the pupils) expressions (statements) and with their justifications for saying so, that is, we are concerned with their *production of knowledge*. Our goal is to get students to operate on those expressions (transform them) first on the basis of a *logic of operations* that has to do with water tanks, water and buckets and then detach, from this *logic*, the transformation rules (*superficially*) produced (that is, in our terms, move it to another semantic field).

Let's agree on calling the amount of water on the left tank X, calling the amount of water on the right tank Y (both 6th grade students' suggestion) and calling the bucket b (teacher's suggestion). A number of statements emerge:

$$X + 4b = Y$$

"because if you *add* 4 buckets on the left there will be only 5 buckets missing"

$$Y - 4b = X$$

"because if you *remove* 4 buckets from the right there will also be 9 buckets missing"

$$X + 2b = Y - 2b$$

"because 7 buckets will be missing on both sides"

$$X + 5b = Y + 1b \text{ (notice: 1b, not b!)}$$

"4 buckets missing on each side"

$$X - 2b = Y - 6b$$

"11 buckets missing on each side"

And then things start to look more interesting:

$$X + 6b = X + 2b$$

"because, because... Oh, no, it's not X, it's Y...!"
Teacher: "What could make that statement true?"
Student: "If the bucket didn't have a bottom!!"

$$X + 20b = Y + 20b$$

"Anything over 5 and 9 buckets will overflow the tanks!!"
Teacher: "$X - 50b = Y - 54b$"
Students: "It's impossible to do that!! You can see from the drawing..."

$$Y - X = 4b$$

"Because if you *remove* from Y the same amount that's in X..."
But how can one *do* that?

This last statement is of special interest. $Y - X = 4b$, $X + 4b = Y$ and $Y - 4b = X$ form the core of a whole-part relationship. Nevertheless, although the *logic of operations* of water tanks, water and buckets renders $X + 4b = Y$ and $Y - 4b = X$ immediately clear and understandable, $Y - X = 4b$ requires some extra imagination: how can one remove X of water from Y if one does not know how much water is in X? One possibility is to remove, say, a glass of water from the left then one from the right, until all the water on the left is removed. Tricky.

It is possible the reader will think I am pushing it too far, but I assure you I am not: we have not met a single instance of spontaneous production of $Y - X = 4b$ by pupils, while $X + 4b = Y$ and $Y - 4b = X$ were always in the first two or three offered. But as soon as the teacher presented them with $Y - X = 4b$ they accepted it as true: "the *difference* between Y and X is 4 buckets" (*of course,* just look at the picture).

Quickly pupils produce a big number of expressions that mean something *to them,* are *legitimate for them,* so now we can begin to operate on those expressions *as objects.*

Teacher: "How would you *describe* the direct transformation[10] of
$X + 4b = Y$ into $X + 5b = Y + 1b$?"
Students: "Add 1 bucket to each tank..."

And from there, with a new *legitimate* action, we could move to generating new legitimate statements offering both "tank-based" and "direct transformation" justifications: $Y - 5b = X - 1b$ either "because 6 buckets missing on each side" or from "remove 1 bucket from each side of $Y - 4b = X$."

[10] Naming of action.

Finally, let's move to *source, target, transformation*: "Transform X + 4b = Y to make it *look like* b = ...".[11] That was the homework. The next day, there were two kinds of answers:

$$b = \frac{Y - X}{4}$$

and

$$1b = Y - X - 3b$$

There was nothing in the whole process that the pupils were not able to do before it. What have they learned, then? They have *learned* that they may do all that *in mathematics*: legitimacy. They have *learned* that expressions may be directly transformed following rules that legitimately (for them) apply to those (legitimate) expressions. They have *learned* that transforming an expression to give it another, given, form is something people do in some situations (legitimacy).

CLOSING REMARKS

I would like to go back to the metaphor I mentioned at the beginning of this paper: if one dives into a lake, before reaching deep (and indeed even shallow) waters, it is necessary to go through the surface. The surface is not an unwanted feature of the depth. Quite on the contrary, it not only delineates where depth is to be found, it is also *the* access gate to depth.

As I said also at the beginning, yes, algebra education must aim at the appreciation and understanding of deeper structures. But what is wrong with mastering superficialities before mastering deep structure? Nothing, I say. But *mastering superficialities* is not the same as *mastering deep structure*, so the former has to be dealt with in its own terms and, I argue, the *superficial* features of mathematics are more of the order of cultural values than of the order of *subject matter*: it is less a matter of *learning* in the more traditional sense and more a matter of *acceptance*, so, *the earlier, the better*.

Unfortunately, too many people are put off by mathematics. But they are not put off by its depth but rather by its surface. That means, I think, that the failure of our students in algebra is not the failure of *those who tried and failed*. It is rather the failure of *those who never tried to succeed in it* (because what they *see* is not legitimate; it does not make sense).[12]

---

[11] *Form*: superficial.

[12] Elsewhere (Lins, 1999) I have published a paper in Portuguese, in which I borrow ideas from Monster Theory (Cultural Studies) to further explore this phenomenon.

The approach we have proposed here, with its explicit attention to an early (yet *superficial*) immersion of pupils into superficial features of algebraic activity, aims precisely, in its broader sense, to give them a chance to really face deep structures. If and when that happens, they will be actually able to decide whether or not they like it. They may decide to eat the spinach; they may decide they prefer pizza.

And that will, ultimately, give *teachers* a chance to teach.

REFERENCES

Bishop, A. (1994, July). Educating the mathematical enculurators. Paper presented at ICMI Regional Conference, Shanghai, China.

Lins, R. (1992). *A framework for understanding what algebraic thinking is.* Unpublished PhD thesis; Shell Centre for Mathematical Education, UK.

Lins, R. (1999). Por que discutir teoria do conhechimento é relevante para a Educação Matemática: Concepções e perspectives. São Paulo: Editora UNESP.

Lins, R. (2001). The production of meaning for algebra: A perspective based on a theoretical model of semantic fields. In R. Sutherland, T. Rojano, A. Bell, & R. Lins (Eds.), *Perspectives on school algebra* (pp. 37–60). Dordrecht: Kluwer Academic.

Lins, R., & Kaput, J. (2004). The early development of algebraic reasoning: The current state of the field. In K. Stacey, H. Chick, & M. Kendal (Eds.), *The future of the teaching and the learning of algebra, the 12th ICMI Study* (pp. 47–70). Dordrecht: Kluwer Academic.

Watson, A. (2008, March). School mathematics as a special kind of mathematics. Paper presented at the ICMI Symposium on the Occasion of the 100th Anniversary of ICMI, Working Group 1 (Disciplinary Mathematics and School Mathematics), Rome, Italy.

Wood, T., Staples, M., Larsen, S., & Marrongelle, K. (2008, March). Why are disciplinary practices in mathematics important as learning practices in school mathematics? Paper presented at the ICMI Symposium on the Occasion of the 100th Anniversary of ICMI, Working Group 1 (Disciplinary Mathematics and School Mathematics), Rome, Italy.

CHAPTER 4

EARLY ALGEBRA

Maria L. Blanton

TEACHING AND LEARNING ALGEBRA IN THE 21ST CENTURY

The teaching and learning of algebra has undergone a critical transformation in the US over the last two decades. With the view that historical paths to algebra have been largely unsuccessful in terms of student achievement and in students' ability to compete internationally in standard mathematics assessments (Hiebert, et al, 2005; Stigler, et al, 1999; US Dept of Education and National Center for Education Statistics, 1998a; 1998b; 1998c), numerous conferences, panels, and study groups have convened to examine the challenges of traditional pathways to teaching and learning algebra and to identify possible solutions. Among others, the US DoE *Algebra Initiative Colloquium* (1993), the NCTM/MSEB *Nature and Role of Algebra in the K–14 Curriculum* Conference (1998) held jointly by the National Council of Teachers of Mathematics (NCTM) and the Mathematical Sciences Education Board (MSEB), the Mathematics Learning Committee of the National Research Council, the International Commission on Mathematical Instruction Study on the Teaching and Learning of Algebra (2001) and, recently, the NSF-funded Mathematical Association of America conference *Algebra: Gateway to a Technological Future* (2006), have worked specifically to formulate recom-

Future Curricular Trends in School Algebra and Geometry: Proceedings of a Conference. pages 45–61

mendations for teaching and learning algebra for this century. As a result of these and other initiatives, scholars have increasingly advocated that algebra be re-conceptualized in school mathematics as a longitudinal strand of thinking that spans grades K–12 (e.g., Kilpatrick, Swafford & Findell, 2001; NCTM, 1989, 2000). A central argument for this shift in thinking is that the culturally-inherited "arithmetic-then-algebra" approach, where 6–8 years of arithmetic instruction in the elementary grades is followed by an abrupt and largely superficial introduction of algebra in secondary grades, has not allowed the time and space for the necessary depth of development in students' mathematical thinking, and as a result, has led to a deep and widespread marginalization of students in school and society (Kaput, 2008). The critical importance of student success in algebra is clear. As Schoenfeld (1995) described,

> Algebra has become an academic passport for passage into virtually every avenue of the job market and every street of schooling. With too few exceptions, students who do not study algebra are therefore relegated to menial jobs and are unable often to even undertake training programs for jobs in which they might be interested. They are sorted out of the opportunities to become productive citizens in our society.

In response to the growing concerns addressed by these and other algebra initiatives, mathematics educators now argue that students need long-term, sustained algebra experiences in school mathematics, *beginning in the elementary grades*, that leverage their natural intuitions about patterns, relationships and structure, and build these ideas over time from informal expressions to more sophisticated and formalized ways of mathematical thinking.

RESULTS OF EARLY ALGEBRA RESEARCH

While much progress has been made in recent years regarding algebra as a K–12 enterprise, the MAA report (Katz, 2007) for the *Algebra: Gateway to a Technological Future* conference advocates that algebra in the elementary grades (hereinafter, *early algebra*) is the most critical need in changing students' success in algebra. Indeed, there is a growing body of research on early algebra regarding what it entails, how elementary students understand algebraic concepts, and what it takes for teachers to create classrooms that build children's algebraic thinking skills. This research has provided us with important existence proofs regarding the capacity of students from diverse socioeconomic and educational backgrounds to think algebraically, including their ability to develop an algebraic, relational view of equality and operate syntactically on equivalent quantities; to identify and symbolize mathematical properties and structure; to transition from natural language to more conventional symbolic algebraic notation in their expressions of generalizations; to use appropriate representational tools—as early as first grade—in the exploration of functional relationships in data; to identify and symbolize functional relationships; to build mathematical arguments that progress from empirical, case-based reasoning to more general arguments that use sophisticated mathematical thinking, and to reason about abstract quantities of physical measures (e.g., length, area, volume), in order to develop algebraic relationships (see, e.g., Bastable & Schifter, 2008; Blanton, 2008; Blanton & Kaput, 2004; Carpenter, Franke & Levi, 2003;Carraher, Schliemann, Brizuela & Earnest, 2006; Dougherty, 2003; Kaput, Carraher & Blanton, 2008; Kieran, 1981; Schifter, 1999)[1].

Not only has this research base challenged the developmental constraints previously placed on young children, it has brought to light the nature of teachers' algebraic knowledge, the forms of professional development needed to transform teachers' algebraic and pedagogical knowledge, and the importance of systemic approaches in creating sustainable early algebra classrooms (see e.g.,; Blanton & Kaput 2005a, 2005b; Blanton & Kaput 2008; Carpenter, et al 2004; Doerr, 2004; Franke, Carpenter, & Battey, 2008; Kaput & Blanton, 2005; Schifter, Bastable & Russell, in press; Stein, Baxter & Leinhardt, 1990).

OUR PARTICULAR RESEARCH FOCUS AND HOW THIS HAS GUIDED CURRICULUM DECISIONS

While it is difficult to define (early) algebra because of variations in the culture of algebra across grades, communities, and even countries, scholars in early algebra research generally agree that algebraic reasoning can be characterized by two core aspects: (1) generalizing mathematical structure, properties, and relationships and expressing these generalizations through increasingly symbolic forms and (2) syntactically-guided reasoning and actions on generalizations (Kaput, 2008). Although there are different views on how these core aspects play out in learning algebra, much of early algebra research is organized around the view that these core aspects can co-emerge when, at the onset of early algebra learning "students are encouraged to note regularities and make generalizations using their own resources, but they are soon encouraged to make conventional representational forms their own" (Kaput, 2008, p. 12). The argument is that conventional forms (e.g., algebraic notation, graphs, tables) can both serve

[1]The recent volume *Algebra in the Early Grades* (Kaput, Carraher & Blanton, 2008) contains the historical development of early algebra research and an overview of results from the field.

as forms of expression and play mediating roles in children's algebraic reasoning.

Although these core aspects represent a critical way of thinking across *all* areas of mathematics, much of the particular curricular content in early algebra research has crystallized around two fundamental domains: (1) the use of arithmetic as a domain for expressing and formalizing generalizations (*generalized arithmetic*) and (2) generalizing numerical or geometric patterns to describe functional relationships (*functional thinking*)[2]. The core aspects of early algebra—generalizing, expressing generality, and reasoning with generalizations—serve as a lens for interpreting the nature of children's algebraic thinking in these two fundamental content domains. Whether students are generalizing arithmetic properties or thinking about how two quantities co-vary, the heart of that activity (and what a researcher might want to know about it) relates to the nature of the generalization being made, its form of expression (e.g., whether conventional algebraic notation or natural language), and how students reason with it.

PROGRESSIONS IN CHILDREN'S FUNCTIONAL THINKING AND ITS IMPLICATIONS FOR CURRICULAR DECISIONS

Within these two curricular domains, much of our research has focused on functional thinking in the elementary grades and our findings have, subsequently, guided the curricular decisions made within these content domains. In particular, we have found that the types of representations children use; the progression of symbolic, notational language in their descriptions of varying quantities and functional relationships; the ways they track and organize data, and their ability to express co-variation and correspondence among quantities (as opposed to just recursive patterning) can be scaffolded in instruction as early as first grade (Blanton & Kaput, 2005b).

Development of recursive, co-variational, and correspondence understanding. Because of this, our research has shaped the nature and complexity of functional thinking tasks we use with elementary students. For instance, beginning in first grade, we make the independent variable explicit in functional thinking tasks, even though students might initially focus on recursive patterns in the dependent variable. We have found that this creates opportunity for children to think about co-variation and correspondence relationships – both critical ideas in the development of functional thinking—and that, with such opportunities, they are able to identify these types of rela-

[2]This assumes early algebra content builds from the arithmetic base typically used in elementary grades. There is also promising early algebra research that does not use arithmetic as a starting point and is not rooted in these two content domains (see, e.g., Dougherty, 2003).

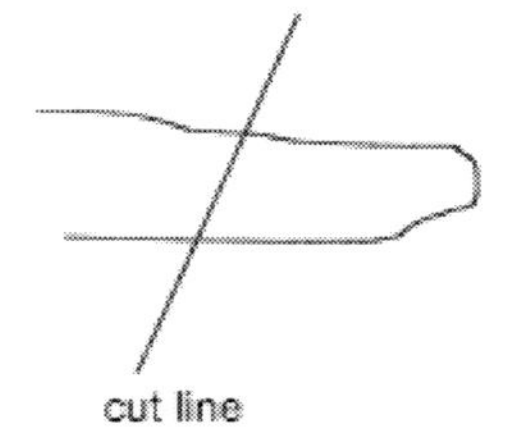

Fold a piece of string to make one loop. While it is folded, make 1 cut (see figure). How many pieces of string do you have? Fold another piece of string to make one loop. Make 2 cuts and find the number of pieces of string. Repeat for 3, 4, and 5 cuts.

FIGURE 1. Cutting a one-loop string.

tionships in primitive ways in the early elementary grades. For example, in a simple task where first-graders are looking for a relationship between the total number of dog eyes for a group of dogs, they can describe the relationship as "it doubles", a precursor to a correspondence relationship such as $E = 2d$, where E represents the number of eyes and d the number of dogs. When looking for a relationship between the number of cuts made to a folded piece of string and the number of pieces of string that result from the cuts, (see Figure 1) they can describe a co-variational relationship of "every time you make one more snip, you get two more pieces of string than you had before". Both of these represent the critical thinking of either explicitly or implicitly attending to how two quantities vary in relation to each other. As early as third grade, children are able to identify and describe correspondence relationships in linear and quadratic functions using algebraic notation. Thus, helping children in early elementary grades move beyond recursive patterning is an important aspect in the development of their functional thinking, and keeping the independent variable explicit in functional thinking tasks is a conscious curricular choice we make to support this transition. Indeed, we have found that recursive patterning, co-variational thinking, and correspondence relationships can co-emerge in young children's thinking. A curriculum that focuses only on recursive patterning in the early elementary grades (as "patterning" tasks have sometimes done) not only misses the opportunity for this to occur, it also risks building a singular way of thinking about patterns in data (recursively) that needs to be "unlearned" in later years.

Development of representational thinking. We have also found that, as early as first grade, students can build representational tools (e.g., function tables, algebraic notation) to support their thinking about functions. This has led to our curricular choices to include these types of representations in the early elementary grades. For instance, the use of function tables as a type of organizing tool has important consequences in the development of children's functional thinking. As noted elsewhere,

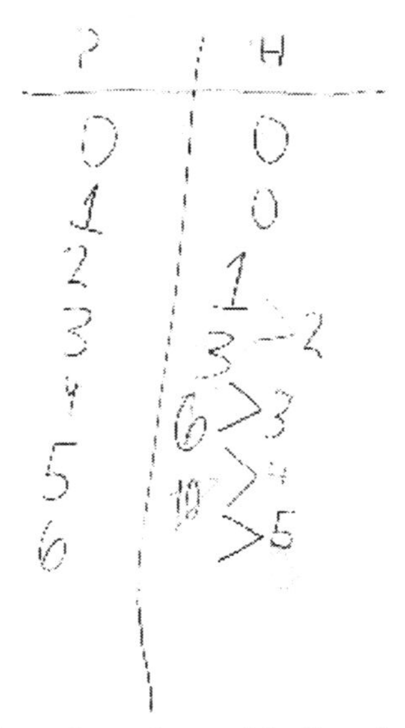

FIGURE 2. A first-grader's function table for the Handshake Problem.

> as children record numerals in a table, they also begin to work on ideas of correspondence between quantities because they are attending to *where* these numbers go in the table and the meanings embodied in the position of the numbers. They learn that certain numbers go in the first column and others in the second column. They learn that when they record a '2' for the number of dogs, they also record '4' for the number of dog eyes in the corresponding position in the next column. This process helps children begin to visually and cognitively look across columns and keep track of two quantities simultaneously, an important early step in functional thinking. (Blanton, 2008, p. 40)

Figure 2 shows a first-grader's function table for the Handshake Problem, based on counting handshakes for a varying number of people in a group (see Blanton, 2008). By first and second grade, the teacher's role in the use of function tables becomes less visible as children begin to recognize where and how to use this tool to organize data. Children become the recorders. Moreover, as they develop meaning for this tool, they are able to shift their focus from issues of numeracy to relationships in the data. As the following teacher reflection on students solving the Growing Snake Problem (see Figure 3) conveys, by third grade, children can use function tables to actively reason about functional relationships:

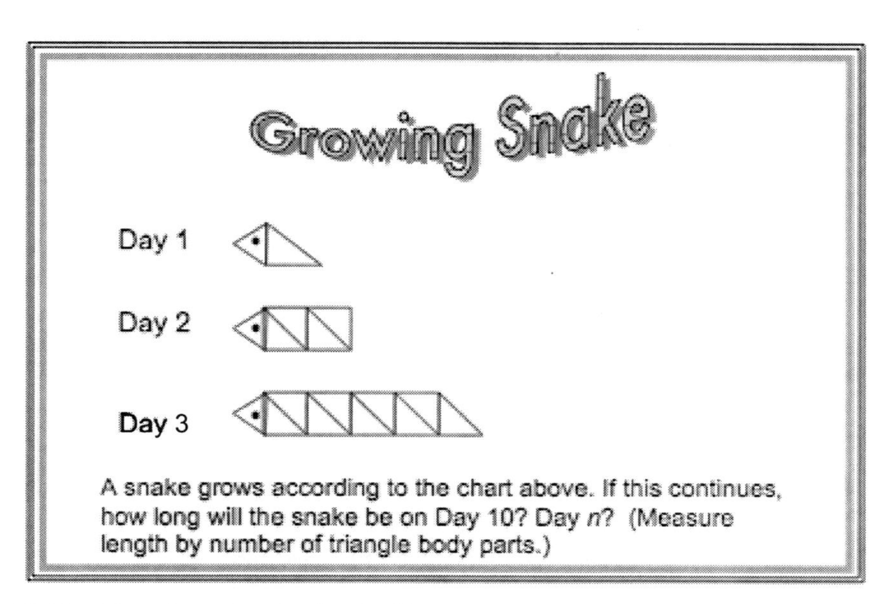

FIGURE 3. Growing Snake Problem.

> "The class worked on this problem for approximately 10 minutes. All organized their data with a t-chart. When I pulled the group together to discuss the problem, it was Karlie who had her hand waving hard.... Karlie usually just sits and listens during math time, so her enthusiasm was very special. I called on her right away. "I know that on day 10 the snake will have 101 body parts and I know that on day n the snake will have '$n * n + 1$'. I know this because I used my t-chart and I looked for the relationship between n and body parts. This is the first time I saw the pattern, so please tell me I'm right!" she said (Blanton & Kaput, 2005b).

We have found that functional thinking tasks also provide an important way to scaffold children's symbol sense and their use of algebraic notation. When curriculum and instruction provide opportunity for children to develop functional thinking, students can transition linguistically from iconic and natural language registers at grades PreK–1 to increasingly symbolic notational systems as early as first grade (Blanton & Kaput, 2004). Fran Vincent wrote about how her first-graders used symbols to represent varying quantities in the Handshake Problem:

> I asked, "Can I label one side [of the t-chart] 'people' and the other side 'handshake'?" One little boy said, "Just write 'p' and 'h'." I immediately stopped what I was doing. I asked, "What did you say?" He continued to re-

> peat what I heard him say. "Awesome, how did you come up with that?" I probed. He continued, "Well, 'people' begins with p and 'handshakes' begin with h" (Blanton, 2008, p. 43).

While the traditional view that development precedes learning might make the use of symbols, as variables, in early elementary grades seem controversial, the perspective we (and many early algebra researchers) take is that learning *promotes* development and that symbolic notational systems are more fully conceptually formed in children's thinking as a result of children's interaction with them in meaningful contexts. In other words, children develop symbol sense as they have opportunity to use symbolic notation in meaningful ways (see also Brizuela, Carraher, & Schliemann, 2000). This has led to curricular choices to introduce symbols as representing varying quantities in the early elementary grades.

DEVELOPING TEACHERS' UNDERSTANDING OF EARLY ALGEBRA

Our work has also fundamentally involved understanding how teachers can build classrooms that develop children's algebraic thinking (Blanton, 2008; Blanton & Kaput, 2005a). This focus has, in turn, affected our curricular decisions about the design of teacher professional development.

Transforming teachers' existing materials and resources. A critical aspect of teacher learning is developing teachers' ability to transform their existing materials and resources into opportunities for algebraic thinking. While new and innovative curricula have begun to include (more) algebraic thinking, many teachers might find that their curricular resources do not address these important ideas. We help teachers look for ways to transform their existing materials, whether by processes such as varying a task parameter or making known quantities unknown (Blanton, 2008; Blanton & Kaput, 2004), so that (typically) arithmetic can become a springboard for a study of generalized arithmetic or functional thinking. As such, our curricular focus in teacher professional development involves helping teachers to find, design, or adapt tasks that fit within their daily curricular plans. Not only does this empower teachers who might otherwise be limited by the curriculum in place in their school and district, we have found that their participation in the development of algebra-focused resources also deepens their knowledge of what early algebra entails. As one first-grade teacher wrote concerning the use of tasks we brought to professional development as well as those she created,

> These activities at the beginning seemed like they were going to be hard to do, never mind creating my own. I've realized that they are a lot simpler to

> create and implement than I thought. I am really impressed with how these activities have shaped my way and my students' way of thinking algebraically. They have really opened my mind up about algebra and how, if we put it into a simple form, our students can do it!

An implicit component of this approach is that, because activities are embedded in teachers' daily practices, it conveys the idea that early algebra is not an "add-on" to the curriculum in place, but a means to make that curriculum deeper and more powerful by extending an emphasis on the particular to one of finding and expressing generality. Through her experiences with this approach to transforming existing materials, one third-grade teacher explained, "Many topics that I cover as part of my curriculum are embedded in algebraic thinking tasks. Therefore, [these tasks] do not have to be considered an add-on, but can be thought of as an extension of what we already do." And, finally, this approach helps teachers develop sensitivity for algebra opportunities in their own classrooms so that their learning is not bound by a particular professional development program and its tightly-specified curriculum. It is critical that teachers' mathematical knowledge and knowledge of practice be generative and self-sustaining so that their learning can extend beyond the life of a particular professional development program.

Developing a school culture of algebraic thinking. A second dimension of teacher learning in our work has involved developing a school culture of algebraic thinking. This occurs both within the classroom as teachers identify ways to integrate algebra across the elementary curriculum and outside of the classroom, as teachers identify ways to focus school attention on algebra ideas. Teachers' abilities to find opportunities for integrating early algebra tasks into other subject areas, whether language arts, social studies, science, or even physical education, give children important experiences with seeing math in action and for building connections across domains of knowledge. Experience with designing opportunities for cross-curricular integration is also an important learning experience for teachers as they make choices about task ideas, relevance of concepts, and modes of presentation. For example, a third-grade teacher designed a functional-thinking task that she used as part of units in social studies, science, and language arts (Blanton, 2008). A fourth-grade teacher developed a functional-thinking task on speed stacking that his students solved during gym class. This not only engages students in algebraic thinking in novel ways, it also pushes teachers' creative thinking about early algebra and leverages their own curricular strengths outside of mathematics.

The creative understanding teachers bring to early algebra is strengthened as they look for ways to increase their school's (or even district's) focus on the development of children's algebraic thinking. Through activities such as implementing a school-wide algebra task in which all classes can participate, designing a "math night" for parents to introduce them to early algebra, or developing a mentoring project pairing students in different elementary grades (Blanton, 2008), teachers learn important aspects of early algebra implementation, such as how students across different grades reason about early algebra ideas and how instruction at a particular grade level might support that learning. As teachers engage in these types of design activities they develop a deeper mathematical knowledge and a deeper sense of how that knowledge is connected to students' experience in the elementary grades.

IMPLICATIONS FOR THE DEVELOPMENT OF MATHEMATICAL REASONING AND PROOF IN STUDENTS' THINKING

Regarding natural tendencies in young children's thinking, Mason (2008) argues that children's capacity to generalize, to identify structure and relationships within the details of the particular, is present from birth and that it is the obligation of formal schooling to cultivate these nascent abilities for mathematical purposes. This perspective in early algebra suggests an important connection to the development of mathematical reasoning and proof in children's thinking. That is, young children bring to formal schooling an innate desire to prove – to convince and to need convincing – that can be cultivated in their mathematical thinking. Early algebra provides a natural context to develop these skills (Maher, in press). In particular, because algebraic reasoning has the inherent goal of developing and justifying mathematical generalizations (see Kaput, Carraher & Blanton, 2008), it can motivate the study of reasoning and proof in the elementary grades. Thus, advocating the development of children's algebraic reasoning as a point of reform (NCTM, 2000) mutually supports the development of children's understanding of proof.

We have found in our research that, while children's arguments typically begin with empirical methods in which they test examples to "prove" a conjecture, children can learn to develop more sophisticated arguments by reasoning with previously-established generalizations, or reasoning from the context of the task and connecting its features and parameters to a conjectured relationship. For instance, third graders were able to construct a more general argument when they reasoned that the sum of three odd numbers was odd because "two odds make an even and when you add odd plus even, you get odd". Their reasoning involved the use of two previously established generalizations ("the sum of any two odds is always even" and "the sum of any odd and any even is always odd") and did not rely solely on an empirical argument of testing sums of three particular odd numbers

trapezoid table problem

Suppose you could seat 5 people at a table shaped like a trapezoid (see Fig. 1).

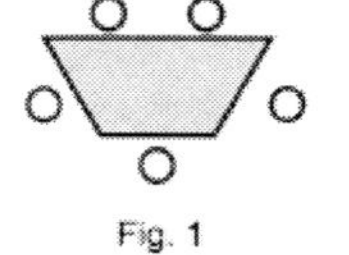

Fig. 1

If you joined two trapezoid tables end-to-end (see Fig. 2) how many people could you seat at the new table? What if you joined 3 trapezoid tables end-to-end? Four tables?

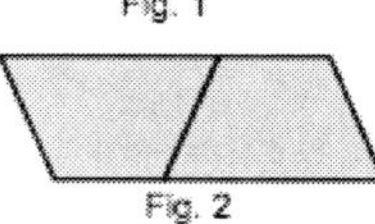

Fig. 2

Organize your data in a table. Can you find a relationship between the number of tables and the number of people seated? Use this relationship to predict the number of people that could be seated at 20 tables and 100 tables. Describe the number of people who could be seated at t tables.

How do you know your relationship works?

FIGURE 4. Trapezoid Table Problem.

number of trapezoids varied, the number of people who could be seated on the sides depended on this varying quantity and could be expressed as "3 times the number of tables", or $3t$. However, the number of people on the ends was always 2 and was always added to the number of people who could be seated on the sides. June wrote, "They realized the 3 came from the people that could sit 'on the top and the bottom' and the 2 came from the two sides" (See Blanton, 2008.)

IMPLICATIONS FOR PRE- AND IN-SERVICE TEACHER EDUCATION

Elementary teachers are in the forefront of early algebra reform, yet have often had little education (as either elementary students or pre-service teachers) or classroom teaching experience with the mathematics of early algebra that they are being asked to teach (Kaput 1999; NCTM & MSEB, 1998). Indeed, we have found that even the expectation of teaching "algebra" in the elementary grades can produce significant anxiety for some elementary teachers. We have also found that, once teachers participate in early algebra professional development connected to their own daily classroom experiences and practices, there are significant shifts in their understanding of (early) algebra and their confidence in being able to teach it. Indeed, some of the more exemplary teachers we have worked with have often described themselves as being weak in mathematics.

All of this suggests that there is real work to be done in both pre-service and in-service elementary teacher education to give teachers the necessary knowledge of both mathematics and teaching mathematics to build classrooms that foster children's algebraic thinking. As Katz argues in reference to K–12 algebra reform, a "massive and sustained effort" will be required to institutionalize early algebra as part of the daily practice of elementary teachers (2007, p. 6). Requiring undergraduate and graduate courses that focus on early algebraic thinking—and how classroom practice supports this—as well as systemic teacher professional development that helps in-service teachers develop the tools and knowledge for building early algebra classrooms is a starting point. More recent ideas involving the development of elementary school specialists who focus on math and science can be another means to focus attention on teacher learning. But the process of institutionalizing early algebra in teachers' daily practice extends beyond course work in an undergraduate degree program or teacher professional development per se, although these are both necessary conditions. The next section explores the real commitment of long-term political, intellectual and economic support from the larger K–16 educational enterprise required to institutionalize early algebra.

PAST, PRESENT, AND FUTURE TREATMENT OF EARLY ALGEBRA IN SCHOOLS: SHIFTS FORWARD, SHIFTS BACK

The last two decades have produced important agendas such as the NCTM *Curriculum and Evaluation Standards* (1989) and the *Principles and Standards for School Mathematics* (2000) that have argued for an increasing emphasis on algebra in the elementary grades as part of a longitudinal K–12 approach to teaching and learning algebra. In our work in Massachusetts, this has filtered down into the development of the Massachusetts Curriculum Frameworks and even the design of the high-stakes MCAS, an accountability assessment for students in most grades K–12. Both the Curriculum Frameworks and MCAS have been instrumental in bringing attention to the K–12 approach to algebra advocated by the NCTM *Standards*, and both recommend explicit attention to algebraic thinking in the elementary grades. Moreover, the importance of both documents in teachers' daily lives (e.g., instructional plans are often mapped to the Curriculum Frameworks, and MCAS' tracking of student performance by student, class, school and district has led to accountability pressures), has helped leverage our work in schools and strengthen the perspective that early algebra is a critical aspect of young children's mathematical development.

In spite of this, we have found that integrating early algebra in deep, longitudinal and sustained ways across the elementary grades can be complicated. A number of factors—political issues, teachers' educational preparation, school and district mandates and priorities for curriculum and teacher ongoing professional development, and even accountability testing, can support or constrain the work of early algebra in schools. While the rhetoric of early algebra is now part of the discourse of reform (NCTM 1989, 2000; RAND Mathematics Study Panel, 2002) and is increasingly integrated into innovative and even mainstream curricula (e.g., in revisions of *Investigations*, Russell, Tierney, & Mokros, 1995), yet with teachers still lacking the mathematical and pedagogical knowledge advocated by the NCTM *Standards* (Mewborn, 2003), we have found that early algebra is often treated as an enrichment exercise in classrooms where it is marginalized in favor of arithmetic instruction.

As we have found in schools in which we work and as noted earlier here, one of the reasons early algebra does not currently receive adequate attention in the elementary grades is that elementary teachers must play a central role in algebra reform, yet typically have had little experience with the types of early algebra content they will need to teach to do so. What are needed are district-based, school-wide systemic approaches to teacher professional development that target all teachers, in all classrooms, in all schools. A systemic approach is critical. Not only does it provide teachers opportunity to build their understanding of early algebra and the types of classroom practice that make it a viable experience for students through a connected, professional community of practice (Blanton & Kaput, 2008; Franke, Carpenter, & Battey, 2008), it allows for the important cross-grade implementation so that students have sustained experiences in algebraic thinking. We have found that the convergence of issues such as the marginalization of algebraic thinking in classroom practice, the lack of adequate teacher preparation, and the lack of systemic approaches to teacher preparation and early algebra implementation have made it difficult, if not impossible, to provide connected, longitudinal algebra experiences for children in elementary grades.

An additional complication to the early algebra effort has been recent reports such as the Curriculum Focal Points (CFP) and the report of the National Mathematics Advisory Panel, which might be seen as de-emphasizing the importance of algebra in the elementary grades in favor of the development of students' proficiency in arithmetic skills. From a philosophy that might be interpreted as suggesting that students are best prepared for algebra in the later grades through a focus on the development of their arithmetic proficiency in elementary grades, some of the specific recommendations found in these documents for the content that really matters across elementary grades seem antithetical to the premise of early algebra. For example, a Grade 3 Curriculum Focal Point for "Number and Operation and *Algebra*" [emphasis added] is developing understandings of multiplication and division and strategies for basic multiplication facts and related division facts (NCTM, 2007). Moreover, as the National Mathematics Advisory Panel Report (2008) notes, the PreK–8 focus for building foundations for algebra involves developing elementary children's proficiency with whole numbers and fractions, and developing particular aspects of geometry and measurement. These recommendations raise questions as to whether early algebra is treated too implicitly, and as a result, how these recommendations might be translated into teachers' daily classroom practices. Given mathematics education reforms in the last two decades and the particular achievements of research on elementary children's algebraic thinking, we are in an unprecedented position to effect positive change in the landscape of K–5 education in the coming decade. Yet in a context for which arithmetic-based instruction has historically been the default position for mathematics in the elementary grades and with recent initiatives such as the *CFP* and the Panel's report, it remains to be seen which way the curricular algebra pendulum will swing in the elementary grades over the next several years.

REFERENCES

Bastable, V., & Schifter, D. (2008). Classroom stories: Examples of elementary students engaged in early algebra. In J. Kaput, D. Carraher, & M. Blanton (Eds.), *Algebra in the early grades* (pp.165–184). Mahwah, NJ: Lawrence Erlbaum.

Blanton, M. (2008). *Algebra and the elementary classroom: Transforming thinking, Transforming practice.* Portsmouth, NH: Heinemann.

Blanton, M., & Kaput, J. (2004). Design principles for instructional contexts that support students' transition from arithmetic to algebraic reasoning: Elements of task and culture. In R. Nemirovsky, B. Warren, A. Rosebery, & J. Solomon (Eds.), *Everyday matters in science and mathematics* (pp. 211–234). Mahwah, NJ: Lawrence Erlbaum.

Blanton, M., & Kaput, J. (2005a). Characterizing a classroom practice that promotes algebraic reasoning. *Journal for Research in Mathematics Education 36*(5), 412–446.

———. (2005b). Helping elementary teachers build mathematical generality into curriculum and instruction. Invited article in Special Edition on Algebraic Thinking, *Zentralblatt für Didaktik der Mathematik (International Reviews on Mathematical Education)*. Edited by Jinfa Cai and Eric Knuth. Vol. 37 (1), 34–42.

Blanton, M., & Kaput, J. (2008). Building district capacity for teacher development in algebraic reasoning. In J. Kaput, D. Carraher, & M. Blanton (Eds.), *Algebra in the early grades* (pp.361–388). Mahwah, NJ: Lawrence Erlbaum.

Blanton, M., Schifter, D., Inge, V., Lofgren, P., Willis, C., Davis, F., & Confrey, J. (2007). Early algebra. In V. Katz (Ed.), *Algebra: Gateway to a technological future* (pp. 7–14). Washington, DC: The Mathematical Association of America.

Brizuela, B., Carraher, D., & Schliemann, A. (2000). *Mathematical notation to support and further reasoning ("to help me think of something")*. Paper presented at the annual meeting of the National Council of Teachers of Mathematics Research Pre-Session, Chicago, IL.

Carpenter, T., Blanton, M., Cobb, P., Franke, M., Kaput, J., & McClain, K. (2004). *Scaling up innovative practices in mathematics and science.* Research Report of the National Center for Improving Student Learning and Achievement in Mathematics and Science. Retrieved May 21, 2008, from www.wcer.wisc.edu/ncisla/publications/index.html.

Carpenter, T., Franke, M., & Levi, L. (2003). *Thinking mathematically: Integrating arithmetic and algebra in elementary school.* Portsmouth, NH: Heinemann.

Carraher, D., Schliemann, A., Brizuela, B., & Earnest, D. (2006). Arithmetic and algebra in early mathematics education. *Journal for Research in Mathematics Education, 37*(2), 87–115.

Doerr, H. (2004). Teachers' knowledge and the teaching of algebra. In K. Stacey, H. Chick & M. Kendal (Eds.), *The future of the teaching and learning of algebra.* (pp. 267–290). Dordrecht: Kluwer.

Dougherty, B. (2003). Voyaging from theory to practice in learning: Measure Up. In N. Pateman, B. Dougherty, & J. Zilliox (Eds.), *Proceedings of the 27th International Conference for the Psychology of Mathematics Education* (Vol. 1, 17–23). Honolulu, HI: College of Education, University of Hawaii.

Franke, M., Carpenter, T., & Battey, D. (2008). Content matters: Algebraic reasoning in teacher professional development. In J. Kaput, D. Carraher, & M. Blanton (Eds.), *Algebra in the early grades.* (pp. 333–359). Mahwah, NJ: Lawrence Erlbaum.

Hiebert, J., Stigler, J., Jacobs, J., Givin, K., Garnier, H., Smith, M., et al. (2005). Mathematics teaching in the United States today (and tomorrow): Results from the TIMSS 1999 Video Study. Educational Evaluation and Policy Analysis, *27*, 111–132.

Kaput, J. (1999). Teaching and learning a new algebra. In E. Fennema & T. Romberg (Eds.), *Mathematical classrooms that promote understanding* (pp. 133–155). Mahwah, NJ: Lawrence Erlbaum.

Kaput, J. (2008). What is algebra? What is algebraic reasoning? In J. Kaput, D. Carraher, & M. Blanton (Eds.), *Algebra in the early grades* (pp. 5–18). Mahwah, NJ: Lawrence Erlbaum.

Kaput, J., & Blanton, M. (2005). Algebrafying the elementary mathematics experience in a teacher-centered, systemic way. In T. Romberg, T. Carpenter, & F. Dremock (Eds.), *Understanding mathematics and science matters* (pp. 99–125). Mahwah, NJ: Lawrence Erlbaum.

Kaput, J., Carraher, D., & Blanton, M. (2008). *Algebra in the early grades.* Mahwah, NJ: Lawrence Erlbaum.

Katz, V. (Ed.). (2007). *Algebra: Gateway to a technological future.* Washington, DC: Mathematical Association of America.

Kieran, C. (1981).Concepts associated with the equality symbol. *Educational Studies in Mathematics 12*, 317–26.

Kilpatrick, J., Swafford, J., & Findell, B. (Eds.). (2001). *Adding it up: Helping children learn mathematics.* Washington, DC: National Academy Press.

Mason, J. (2008). Making use of children's powers to produce algebraic thinking. In J. Kaput, D. Carraher, & M. Blanton. (Eds.), *Algebra in the early grades* (pp. 57–94). Mahwah, NJ: Lawrence Erlbaum.

Mewborn, D. (2003). Teaching, teachers' knowledge, and their professional development. In J. Kilpatrick, W. G. Martin, & D. Schifter (Eds.), *A research companion to the Principals and Standards for School Mathematics* (pp. 45–52). Reston, VA: NCTM.

National Council of Teachers of Mathematics. (1989). *Curriculum and evaluation standards for school mathematics.* Reston, VA: Author.

National Council of Teachers of Mathematics. (2000). *Principles and standards for school mathematics.* Reston, VA: Author.

National Council of Teachers of Mathematics. (2007). *Curriculum focal points for prekindergarten through grade 8 mathematics.* Reston, VA: Author.

National Council of Teachers of Mathematics & Mathematical Sciences Education Board. (1998). *The nature and role of algebra in the K–14 curriculum: Proceedings of a national symposium.* Washington, DC: National Research Council, National Academy Press.

National Mathematics Advisory Panel Report. Retrieved May 21, 2008, from www.ed.gov/about/bdscomm/list/mathpanel/report/final-report.pdf.

RAND Mathematics Study Panel Report. (2002). *Mathematical proficiency for all students: A strategic research and development program in mathematics education.* Washington, DC: U.S. Department of Education.

Russell, S., Tierney, C., & Mokros, J. (1995). *Curriculum modules in elementary mathematics: Investigations in number, data, and space.* Palo Alto, CA: Dale Seymour Publishing Co.

Russell, S. J., Schifter, D., & Bastable, V. (January/February, 2006). Is it 2 more or 2 less? Algebra in the elementary classroom. *Connect, 19*(3), 1–3.

Schifter, D. (1999). Reasoning about operations: Early algebraic thinking, grades K through 6. In L. Stiff & F. Curio (Eds.). *Mathematical Reasoning, K–12: 1999 NCTM Yearbook.* (pp. 62–81). Reston, VA: NCTM.

Schifter, D., Bastable, V., & Russell, S. (in press). *Developing mathematics ideas casebook, facilitator's guide, and videotape for reasoning algebraically about operations.* Parsippany, NJ: Pearson.

Schoenfeld, A. (1995). Is thinking about 'Algebra' a misdirection? In C. Lacampagne, W. Blair, & J. Kaput (Eds.), *The Algebra Colloquium. Volume 2: Working Group Papers* (pp. 83–86). Washington, DC: US Department of Education, Office of Educational Research and Improvement.

Stein, M. K., Baxter, J., & Leinhardt, G. (1990). Subject-matter knowledge and elementary instruction: A case from functions and graphing. *American Educational Research Journal, 27*(4), 639–663.

Stigler, J. W., Gonzales, P., Kawanaka, T., Knoll, S., & Serrano, A. (1999). *The TIMSS videotape classroom study: Methods and findings from an exploratory research project on eighth-grade mathematics instruction in Germany, Japan, and the United States.*

(NCES1999-074). U.S. Department of Education. Washington, DC: National Center for Education Statistics.

U.S. Department of Education & National Center for Education Statistics. (1998a). *Pursuing excellence: A study of U.S. fourth grade mathematics and science achievement in international context.* Washington, DC: U.S. Government Printing Office.

———. (1998b). *Pursuing excellence: A study of U.S. eighth-grade mathematics and science achievement in international context.* Washington, DC: U.S. Government Printing Office.

———. (1998c). *Pursuing excellence: A study of U.S. 12th-grade mathematics and science achievement in international context.* Washington, DC: U.S. Government Printing Office.

CHAPTER 5

A DAVYDOV APPROACH TO EARLY MATHEMATICS

Barbara J. Dougherty

For many, early mathematics, in grades K–3, should be devoted to number and numeration because it is a time when counting and computational algorithms are developed with some fluency. However, when the focus is on number in this more traditional manner, young children do not develop an understanding of the structure of mathematics or of quantitative relationships in ways that give them access to more complex ideas. This paper presents another way to think about early mathematics so that young children become big thinkers about mathematical properties and structure, using sophisticated ideas without using numbers.

PREMISES AND APPLICATIONS OF THE WORK OF DAVYDOV

Davydov (1975a) suggested in his writings that a different approach from the traditional counting and computational paths might offer children the opportunity to understand mathematics in a more robust way. His writing provokes us to seriously consider a new way to construct mathematics for young children, not just to think about new activities or tasks within number and computation. He was bold enough to suggest that children could

Future Curricular Trends in School Algebra and Geometry: Proceedings of a Conference. pages 63–69

actually learn about and understand mathematical structure—and properties—through quantitative relationships and measurement contexts.

After much reading of his work and the curriculum materials that were developed (Davydov, Gorbov, Mukulina, Savelyeva, & Tabachnikova (1999), and considering the ramifications of what he suggested, I, along with a team of other researchers at the University of Hawaii, decided to investigate his ideas with children. We began this investigation at grade 1, in what Davydov (1975b) referred to as the prenumeric stage. This stage sets the foundation for all of the mathematics that follows in subsequent grades.

In the prenumeric stage, children first identify the attributes of objects that can be compared, or measured so to speak. For example, one book can be compared to another book with regard to their lengths, the areas of their covers, or their masses. Of course, there are other qualities that can be compared but these are three that could be quantified if we wanted.

As the attributes are identified, children needed a way to be able to find and describe the relationships of, say, the lengths of the books. To find their relationships, the lengths could be laid next to each other and directly compared. Now came the dilemma of how we could describe the relationship. Let's take the case that the two lengths were not equal. In order to convey that information, we had to have a way of naming the length of each book. We could call one length B and the other length W. The variables selected are arbitrary and do not represent the object. Instead, the variable represents the attribute being measured or compared.

A natural way of describing the relationship is to say that length B is not the same as length W. Another way is to draw line segments to show the relationship (see figure 1). Yet another way is to use paper strips to show that one is longer than the other. And, finally, we could write statements, such as $W < B$, $B > W$, $B \neq W$, $W \neq B$. By using the multiple representations that include a physical or direct comparison, an intermediate comparison (use of line segments and paper strips), and symbolic language (statements) as well as natural language, children develop mental maps of how quantities

FIGURE 1. Line segments showing comparison of lengths B and W.

can relate. (Note, if the two lengths had been the same, children would have written $B = W$ and $W = B$.)

It is important to recognize that six-year-olds are able to use these literal symbols to represent the relationships of attributes of a continuous nature without any numbers. In the first year of our research on the prenumeric stage, it was surprising to see children fluently use this notation system with the ability to articulate the meaning of the symbols. When I think of the first graders, I am reminded of Justin who told me that he really liked 'doing inequalities' because you could always write four statements. He thought that was more fun than writing only two equality statements!

During this prenumeric stage, children moved from direct comparisons to thinking about issues of equality and inequality. They were presented with tasks that forced them to confront the issues of creating two quantities that were equal from two quantities that had been determined to be unequal. Let's continue to use the two lengths B and W. To make the two quantities equal, children identified the length that needs to be added to length W to create a length as long as length B (see figure 2 for the line segment model). At the same time, they noticed that this is the same length that should be taken away from length B so that the new length would be equal to length W. To notate the actions they did, they recorded statements such as $B - S = W$ or $W + S = B$. It was also possible to write that $B > W$ by S (or $W < B$ by S. Notice that they assigned a name to the length that represented the difference in lengths B and W. This idea that you can add a quantity or subtract the same quantity to create two equal amounts is an important idea that supports later computation development.

You might notice that the take-away model of subtraction, as well as the comparison and missing-addend models, were represented with these tasks, again, without number. It is worthwhile to note that this also represents the joining model of addition. Thus addition and subtraction are represented concurrently so that children see the links between the two computations.

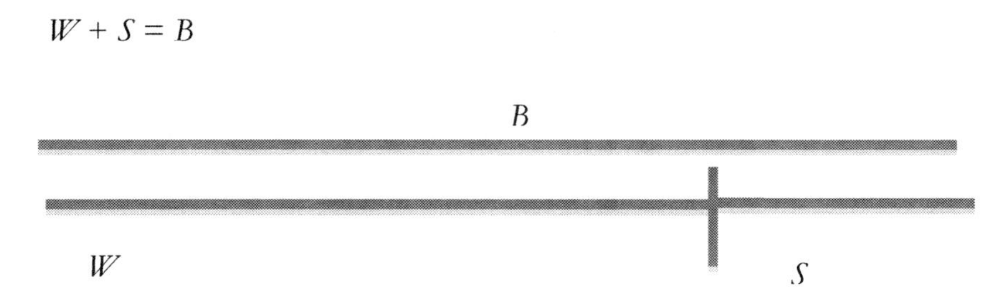

FIGURE 2. Equations and line segments representing how the lengths can be made equal.

These comparisons and associated statements are relatively easy for young children to develop, especially when they can make direct comparisons of length, area, volume or mass. But what happens when two quantities, such as length, cannot be directly compared? To motivate that discussion, children were given two lengths, such as the length of a bookcase and the height of a bookcase across the room. They could not move the two bookcases to make a direct comparison. They were challenged to find a way to decide if the lengths are the same, or if one is longer than the other. They decided to use in intermediary unit. That is, they made the length of one bookcase using string. They then used the string to compare it to the other length (or height). This led quite naturally to an introduction of the transitive property of equality.

The children symbolically represented this as:

$$\begin{array}{c} A = P \\ \underline{P = R} \\ A = R \end{array}$$

What the children said is that if length A and length P are the same, and length P is the same as length R, then length A **has to be** (their emphasis) equal to length R. In this example, length P represented the intermediary unit that was used to compare one length to the other.

What is important to remember here is that the symbolization shown above is used by six-year-old children. Yet, even at this young age, they are able to explicate what it means and model it in a variety of ways, including the physical models, the intermediate representations (line segments), natural language, and symbolically.

The use of these non-specified or general quantities illustrates one of Davydov's critical points about early mathematics. He (1975a) points to the need for students to be able to see the general, rather than the specific, in their early mathematical development. In the above example, children literally see the transitive property of equality in the physical models and represent it in symbolic terms without applying any numerical values (which would be specific cases of this property). This makes use of Vygotsky's (1976) notion of scientific concept development where the instruction begins with general contexts and moves to the specific, rather than the other way around as we would see in a traditional curriculum. From the generalized ideas of the mathematics, children are able to see number as specific cases. The power of this approach is that children do not compartmentalize or fragment mathematical ideas. They see them as connected in a natural way.

The development in grade 1 continues with the introduction of unit, one of the most important pieces of the mathematical foundation children are building (Minskaya, 1975). In a traditional approach, children at this grade level would be counting by rote up to 100 or beyond and counting groups of 10s and 100s of objects in a one-to-one correspondence. Using the Davydov approach, these first graders began their counting experiences with tasks that have students measure out the quantity, whether it is length, area, volume or mass, with different units to explore the effects of choosing different sized units to define the counting. For example, in figure 3, area Z can be measured by an area-unit E that would fit into the area 3 times. A different area-unit H could be used such that it would fit into the area 6 times. Thus the measure or count is different depending upon the unit chosen to measure the quantity.

It is possible for these young children to then write statements that capture the measurement perspective. For example, one statement might be $Z/E = 3$, signifying that some quantity Z (in this case, area), measured by some unit E, was 3. In the case of the area-unit H, the statement would be similar, $Z/H = 6$. Surprisingly, even without the physical representations, six-year-olds can make sense of the statements, and even make comparisons. For example, if no physical representation had been given, children would note that area-unit E must be larger than area-unit H because it took more of the area-unit H to measure area Z than it did for the area-unit E. Of course, this doesn't happen the first time children use or see these symbolic statements, but in a relatively short amount of time, they make sophisticated observations about the relationships across quantities and units, given only the symbolic statements.

What this leads to is the realization by these children that in order to compare measurements or counts of quantities, you have to use the same unit. This, of course, is fundamental to development of rational number concepts, explicit measurement ideas, and many others. To see how this develops involves other milestones along the way that continue to reinforce and build on this idea.

FIGURE 3. Area Z to be measured by different units.

Further, given the children's previous experiences with writing multiple statements about the same relationship, they are quite comfortable now to use the statements above to write other ways in which the area-units and the total quantity relate. For example, let's take the relation of area Z and area-unit H. Other statements would include:

$$6H = Z$$

$$Z = H + H + H + H + H + H$$

$$Z - H = 5H$$

$$Z = H6$$

Others may also be written but rather than supply the exhaustive list of what children might write, consider what these statements represent. The relationship between the unit used for measuring and the whole quantity is quite clear in each of the statements.

The relationship of the quantities can be represented in diagrammatic form as well. Consider a quantity M with parts C and J. The diagram in figure 4 shows how the relationship might be represented. The power of this representation is that the relationship between the whole and the parts is explicit. As you can imagine, many different equations and inequalities can be expressed from this representation, including inequality statements, such as $M > J$ by C. As one might guess, content presented in this way leads naturally to the use of variables to represent unknown amounts and thus finding the value(s) of the unknowns.

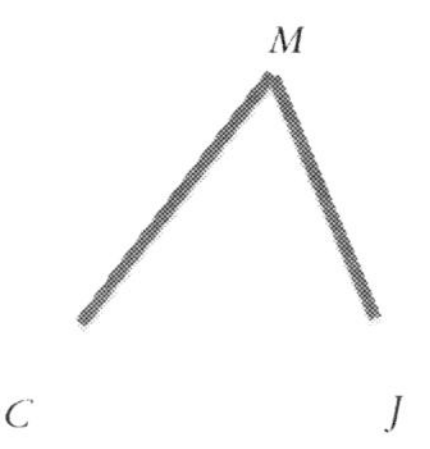

FIGURE 4. Part-whole diagram, showing M as the whole and C and J as the parts.

The development in grade 1 continues with the introduction of number in about seventh month of school. This is a large departure from what would be considered the norm for early mathematics but it is clear that the mathematical foundation that children bring to the number contexts is significantly richer than beginning with specific number tasks.

SUMMARY

The work of Davydov has inspired me to think outside of the box about early mathematics. The ideas represented in his work and implemented with children have shown that very young children are capable of thinking about more sophisticated ideas and of acting on non-specific quantities that are not countable. This approach holds the promise of helping children become comfortable with symbolic representations long before formal algebra is accessible to them.

REFERENCES

Davydov, V. V. (1975a). Logical and psychological problems of elementary mathematics as an academic subject. In L. P. Steffe, (Ed.), *Children's capacity for learning mathematics. Soviet Studies in the Psychology of Learning and Teaching Mathematics, Vol. VII* (pp. 55–107). Chicago: University of Chicago.

Davydov, V. V. (1975b). The psychological characteristics of the "prenumerical" period of mathematics instruction. In L. P. Steffe, (Ed.), *Children's capacity for learning mathematics. Soviet Studies in the Psychology of Learning and Teaching Mathematics, Vol. VII* (pp. 109–205). Chicago: University of Chicago.

Davydov, V. V., Gorbov, S., Mukulina, T., Savelyeva, M., & Tabachnikova, N. (1999). *Mathematics.* Moscow: Moscow Press.

Minskaya, G. I. (1975). Developing the concept of number by means of the relationship of quantities. In L. P. Steffe (Ed.) *Children's capacity for learning mathematics. Soviet Studies in the Psychology of Learning and Teaching Mathematics, Vol. VII* (pp. 207–261). Chicago: University of Chicago.

Vygotsky, L. S. (1978). *Mind in society.* Cambridge, MA: Harvard University Press.

PART II

ALGEBRA USING COMPUTER ALGEBRA SYSTEMS (CAS)

CHAPTER 6

TECHNOLOGY AND THE YIN AND YANG OF TEACHING AND LEARNING MATHEMATICS

Bernhard Kutzler

INTRODUCTION

Humans are ruled by two forces: *Hold on* and *Let go*. These two forces correspond to the two polar energies, the *Yin* and the *Yang*. In the above context, these two energies manifest as *Connect* and *Automate*. *Connecting* is a seeking form of *holding on*. *Automating* means to let a tool do what we used to do ourselves—we *let go* of these tasks.

Hold on	-	Yin	-	Connect
Let go	-	Yang	-	Automate

A car automates transportation. Instead of walking to the shop, we can go there by car. This saves us from having to walk and from carrying the groceries. For some people, using a car for their shopping is a *convenience*.

Future Curricular Trends in School Algebra and Geometry: Proceedings of a Conference. pages 73–91

For people who are physically challenged, using a car for their shopping may be a matter of *survival*.

There are two motivations for automation: *Amplification* and *compensation*. Optical instruments such as telescopes and microscopes *amplify* our natural eyesight so that we can see things that we cannot see otherwise. Optical instruments such as eyeglasses *compensate* poor eyesight so that people with poor eyesight can see things that people with normal eyesight can see without glasses.

Amplification has two aspects based on the motivation to amplify. One can use a telescope to look at a distant object as an astronomer does when observing a lunar eclipse. One can use a telescope to scan the sky in the search for new stars. These two uses may be named *solving* and *exploring*.

This gives three automation archetypes based on the motivation to automate:

Automate = Compensate + Solve + Explore (Figure 1)

Connection also comprises three archetypes based on what to connect with what, notably *representation*, *documentation*, and *communication*. *Representation* is about connecting models with models, such as connecting an algebraic model (an expression) with a graphic model (a graph) or a numeric model (a table). *Documentation* is about connecting models with humans, such as writing a paper on how a problem was solved. *Communication* is about connecting humans with humans, such as having students work in groups.

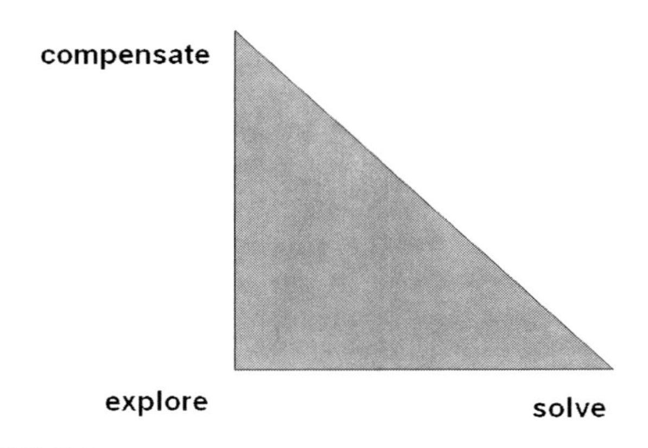

FIGURE 1.

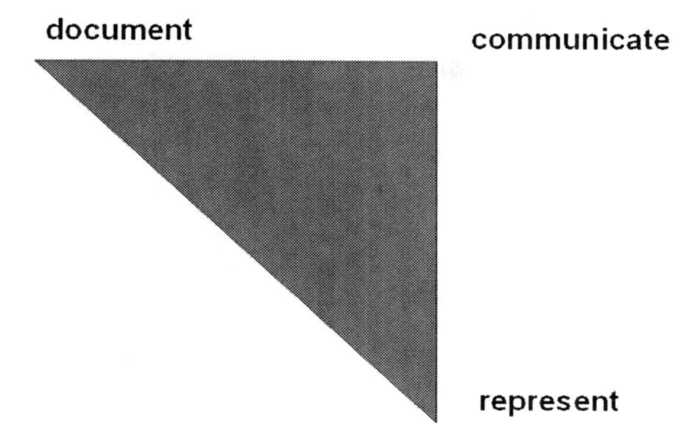

FIGURE 2.

Connect = Represent + Document + Communicate (Figure 2)

Putting these next to each other yields a picture that I call the *Yin and Yang of teaching and learning mathematics* (Figure 3).

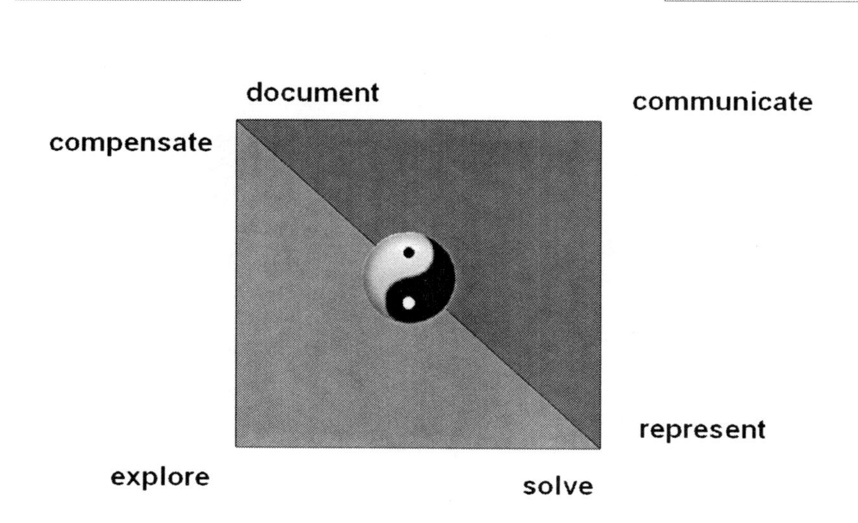

FIGURE 3.

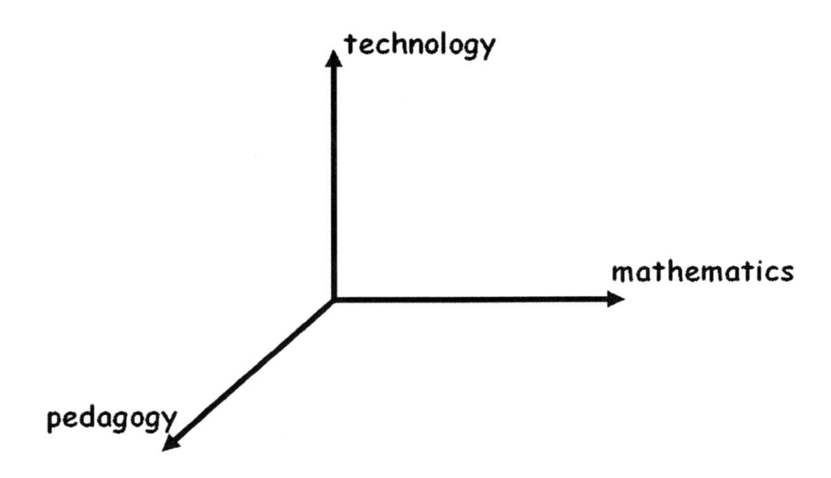

FIGURE 4.

MATHEMATICS/PEDAGOGY/TECHNOLOGY-SPACE

Let the x-axis of 3-D space represent mathematics, the y-axis represent pedagogy, and the z-axis represent technology (Figure 4).

In this model, mathematicians are people who work "on the mathematics axis" and mathematics teachers are people who work "in the mathematics/pedagogy plane."

More and more mathematicians accept computer software as tools for research. These mathematicians move from the 1-dimensional (1-D) mathematics axis into the 2-dimensional (2-D) mathematics/technology plane. The benefit is that one has infinitely many more ways of connecting two points on the mathematics axis by allowing paths in the mathematics/technology plane (Figure 5).

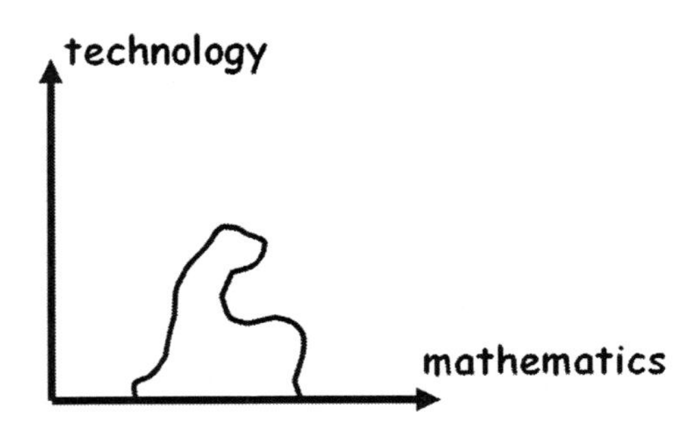

FIGURE 5.

Technology makes some mathematics possible. (Bert Waits)

The same argument holds for accepting computer tools for teaching and learning mathematics. Mathematics teachers using technology move from the 2-D mathematics/pedagogy plane into the 3-D mathematics/pedagogy/technology space, which means infinitely many more possibilities of connecting two points in the mathematics/pedagogy plane by permitting 3-D paths. This is like allowing helicopter trips in a landscape that previously could be explored only using "surface-attached" tools such as bikes, cars, or ships.

Technology makes some mathematics pedagogy possible. (Bert Waits, extended)

REPRESENT

Consider the following problem: A homogeneous cube hangs by a thread attached to one corner. It is otherwise free to move. When we look at this configuration from the front, we see the cube as three parallelograms. What is the smallest angle (greater than zero) through which we must spin the cube so that we see the same figure?

Trivially we get the same figure after rotating the cube a full 360 degrees, but does it happen earlier?

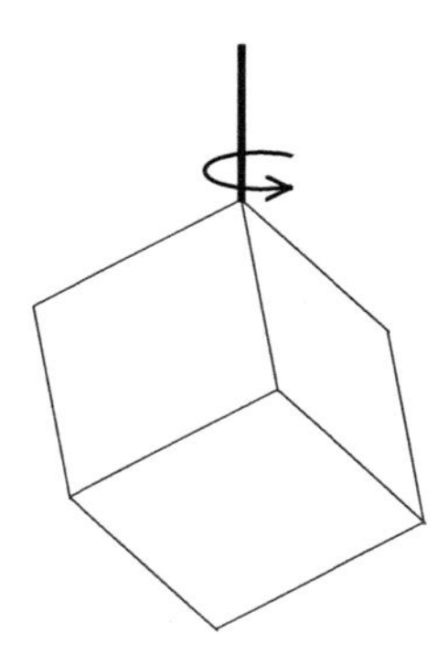

FIGURE 6.

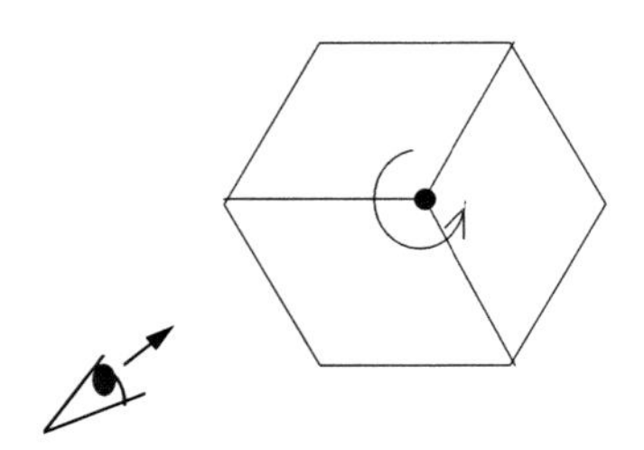

FIGURE 7.

We vary the representation in that we look from above, where the thread attaches (Figure 7).

From this point of view the answer is obvious. For a rotation of 120 degrees, 240 degrees, and 360 degrees the figure remains the same, so 120 degrees is the solution.

The problem appeared demanding when looking at the first picture. The solution is obvious when looking at the second picture.

If you have a problem, there are two paths open to you:
either you solve the problem, or you change your view.
(Chinese Proverb)

Changing the view means changing the representation.

Representations are central in mathematics. One of the basic techniques of mathematical problem solving is to find a representation of a problem that makes the solution obvious. Therefore, one can look at mathematical problem solving as the *art of transforming representations until the solution is visible.* The hanging cube is a fine example. Another example is the solving of:

$$5x - 6 = 2x + 15$$

by transforming it into the equivalent equation $x = 7$.

Figure 8 shows various representations of:

$$x^2 - 2.$$

Double-headed arrows indicate equivalence. Single-headed arrows indicate loss of information.

$$\left(x+\sqrt{2}\right)\cdot\left(x-\sqrt{2}\right)$$

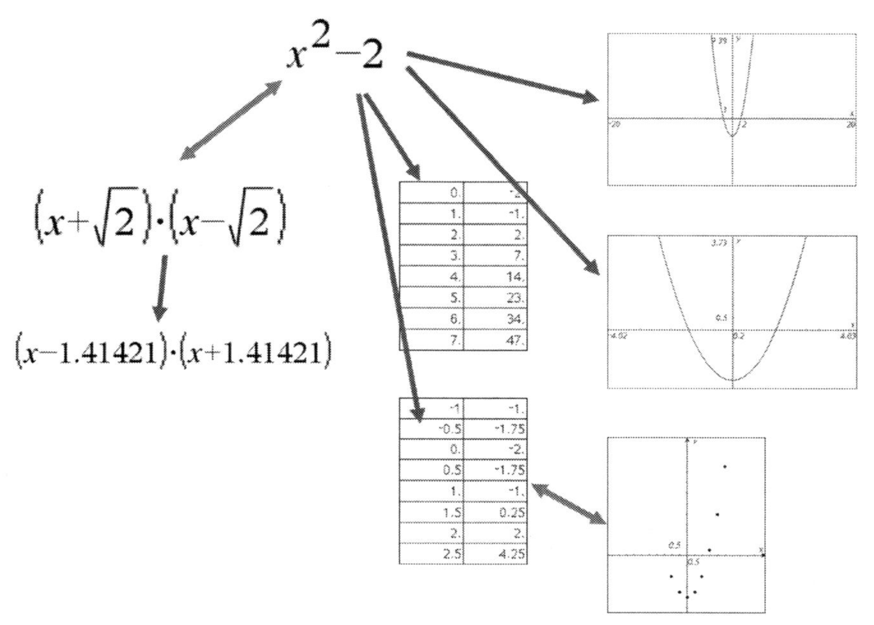

FIGURE 8.

is the factored equivalent of

$$x^2 - 2$$

and one can "go back" by expanding the expression.

$$(x+1.41421)\cdot(x-1.41421)$$

is a decimal approximation. The two tables are function tables for:

$$x^2 - 2$$

with different starting values, increments, and end values. Each table encompasses a significant reduction of information. Graphs are geometric equivalents of function tables obtained via the Cartesian coordinate concept.

One of the core skills of a mathematician is to hold different representations of an object in his mind and to choose the one that is most useful in a given context.

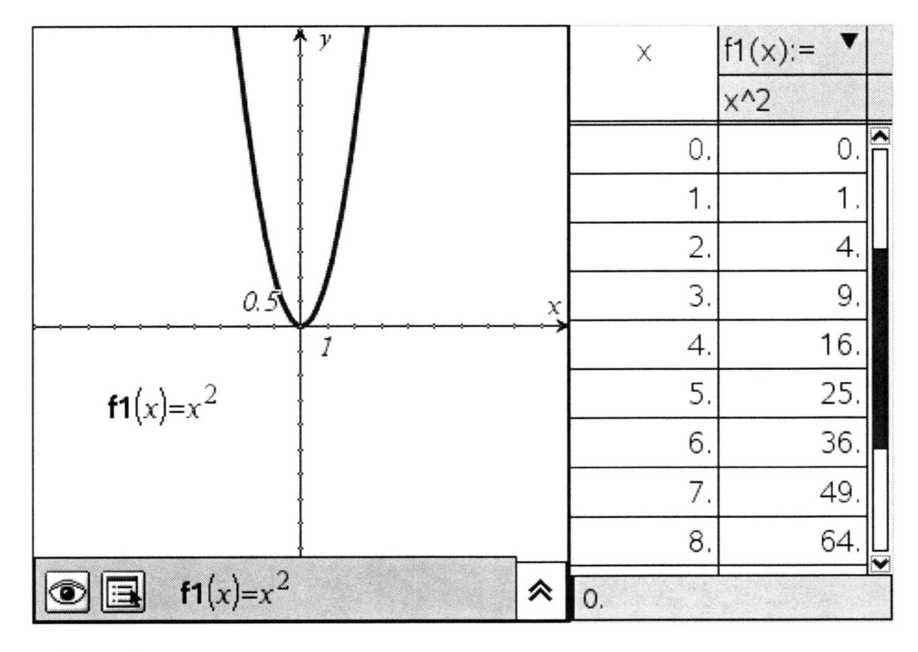

FIGURE 9.

Mathematics teachers strive to help students develop this skill. Students cannot "look into a mathematician's brain," but teachers can employ technology to simulate a mathematician's mind. Seeing several representations of an object right next to each other and seeing how all other representations change when one representation is modified is a powerful pedagogical approach that is possible only with computers.

Figure 9 shows a TI-Nspire screen with the function expression:

$$f1(x) = x^2,$$

a corresponding function graph, and a corresponding function table.

One can edit the function expression, and then observe how the function graph and the function table change. One can grab the function graph, drag it, and then observe how the function expression and the function table change.

DOCUMENT

Documenting requires the ability to *argue*, i.e. to *convince* somebody with a certain level of knowledge. The ability to argue requires the ability to *design* the content (which needs *creativity*) and the ability to *describe*–to convey a message with an intended meaning in an unambiguous manner.

Describing is the inverse of *understanding*, which is the skill to interpret a given natural language text or document. *Argumentation* is about finding convincing reasons to support your conclusions.

In traditional assessments, the solution to the problem is often the only goal and the craftsmanship of performing the calculations that are required to obtain a solution earns a student a good mark. This changes with technology, because finding a solution is a different kind of work when all you need is to press keys. It still is *mathematical* work because choosing the appropriate sequence of keys from a large selection of keys does require mathematical knowhow. However, this process requires less time than performing the underlying calculations with paper and pencil.

The result loses importance because it is easy to obtain. A documentation of how the problem was solved is more than just a replacement: documentation of experimenting or problem solving provides valuable feedback in the teaching process. For assessment, documentation is comparable to a composition written for an English class.

Technology has two roles for the *document* archetype:

1. Technology means a shift of focus from executing algorithms to documenting mathematical work, i.e. it strengthens the role of documentation in the classroom.
2. Technology can support the production of documents.

COMMUNICATE

Communication is the most natural mental activity of humans. It is related to the word *community*, hence reflects a central aspect of the human as a social being. The means of communication is *language*. Natural languages, such as English connect humans with humans. Mathematics connects humans with nature.

The book of nature is written in the language of mathematics.
(Galileo Galilei)

We use communication in the narrow sense as a connection between humans. In fact, *documentation* is an "offline" form of *communication*. Therefore, the skills discussed in the previous section are relevant also here.

First, we look at the communication between student and teacher. Traditionally, teachers use *front teaching*, where the teacher is active and the students passive. The opposite would be a classroom setting in which the students work either freely exploring or under the guidance of the teacher. Such teaching situations encourage the creativity of the students. The ideal lies in a good mixture of these two forms.

Next, we look at the communication between students. Traditionally, students work *alone*, as is most typically enforced in an exam. Later in life they will need to also work in teams, so *teamwork* should be encouraged in school already. A good mathematics teaching and learning tool should support communication, for example through exchanging documents and screen content.

COMPENSATE

Say, we have to solve the equation

$$x + 6 = 18 - 2x$$

This is done by transforming it into the form "x = term with no x" through choosing and applying a sequence of equivalence transformations. Typically one will "*bring terms with x to one side of the equation*" and "*bring all other terms to the other side.*" A good first choice is to add $2x$ to both sides of the equation.

$$x + 6 = 18 - 2x \quad | + 2x$$

After *choosing* this transformation, we have to *apply* it.

$$\begin{aligned} x + 6 &= 18 - 2x \quad && | + 2x \\ 3x + 6 &= 18 && | - 6 \end{aligned}$$

Now it would be appropriate to subtract 6.

$$\begin{aligned} x + 6 &= 18 - 2x \quad && | + 2x \\ 3x + 6 &= 18 && | - 6 \\ 3x &= 12 && | - 3 \end{aligned}$$

We don't have to care about the students who succeed. We should care about the students who don't. Let's find out why they make certain errors and how we can help them to avoid these errors.

Back to the equation $3x = 12$. At this point some students find it difficult to choose a good next step. The following argument is typical: *"There is a 3 in front of the variable x. To get rid of the 3, I must subtract 3."*

A student, who uses this argument, most likely will proceed as follows:

$$
\begin{aligned}
x+6 &= 18-2x \quad &| +2x\\
3x+6 &= 18 \quad &| -6\\
3x &= 12 \quad &| -3\\
x &= 9
\end{aligned}
$$

The student will believe that the equation is solved.

How can technology help to make it better? There are two alternating tasks: (1) the choice of an equivalence transformation and (2) the simplification of algebraic expressions. The choice of an equivalence transformation is a higher-level task. It is the new skill that has to be learned when learning to solve equations. The simplification is a lower-level task, for which the teacher has to assume that the student is sufficiently well trained:

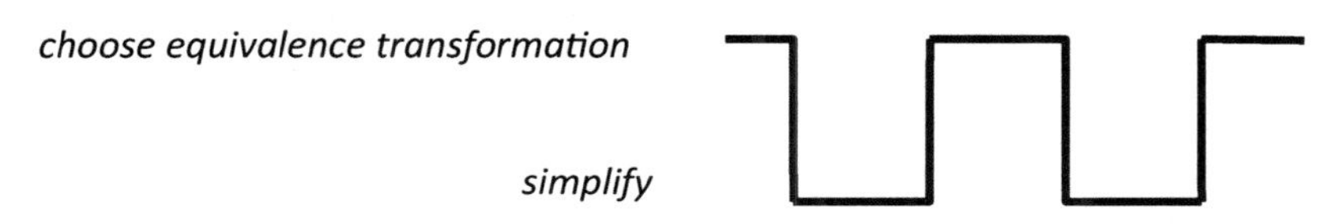

A student, while trying to learn the new skill, repeatedly has to interrupt the learning process in order to perform a simplification. Unfortunately, the interruption can influence the "game:" A mistake made during the simplification severely disturbs the higher-level task and may prevent the student from learning. After deciding to subtract 3, the student should "forget" the motivation for this choice. But, in reality, the student starts the next line with "x =" simply "*because the transformation –3 was chosen in order to generate 'x =' on the left hand side.*" At the higher level, the student has the (wrong) impression that –3 simplified the equation.

In the sequel, we demonstrate how one can use TI-Nspire CAS to help students in this situation.[1]

Enter: $x+6=18-2x$

$x+6=18-2\cdot x$	$x+6=18-2\cdot x$

Enter: $+2x$

$(x+6=18-2\cdot x)+2\cdot x$	$3\cdot x+6=18$

Enter: –6

[1]The original source of this approach is Aspetsberger/Funk, see *References*.

$(3 \cdot x+6=18)-6$ $3 \cdot x=12$

So far everything is like it was with paper and pencil. Now we mimic a student who chooses to subtract 3.

Enter: –3

$(3 \cdot x=12)-3$ $3 \cdot x-3=9$

The tool applies the transformation correctly and the student *immediately* sees that subtracting 3 did *not* simplify the equation as expected. The tool gave important feedback on the quality of the student's choice. It is like putting the finger on a hot stove and feeling the pain *immediately*.

This approach is called the scaffolding method. It is based on what Bruno Buchberger, in the mid 80s, suggested as the "Black-Box-White-Box Principle."

Say we taught and practiced the solving of systems of linear equations. When we have to move to the next topic, some of our students will have mastered systems of linear equations while others will have not. Say the next topic is analytic geometry. Many analytic geometry problems require the solving of systems of linear equations. Students who still struggle with such systems will find it difficult if not impossible to solve most of the analytic geometry problems.

Good eyesight is a prerequisite for driving a car safely. What about people with poor eyesight? Should they be banned from the roads? Fortunately they can use *eyeglasses* that make up for their weakness. Accordingly, we should allow students with a poor solving-systems-of-linear-equations skill to use a compensation tool when "driving in analytic geometry land." This is not only an act of humanity, this is our pedagogical duty!

Analytic geometry should not be (mis)used as a therapeutic opportunity to repair a solving-systems-of-linear-equations weakness!

SOLVE

Traditional mathematics teaching is centred on solving problems—ranging from simple calculations such as $3x + 4x = ?$ to complex word problems involving optimization.

Technology for supporting one of the other five archetypes (*Represent, Document, Communicate, Compensate, Explore*) is mostly seen as an enrichment of traditional teaching. Technology for solving problems, however, often is seen as a threat.

CAS provides a rich collection of black boxes. Therefore CAS polarizes educators into supporters and opponents. Supporters would like to use CAS whenever possible, because this would allow for solving more (realistic) problems. Opponents would like to ban CAS, because they believe that scientific calculators destroyed their students' mental arithmetic and CAS could have an even more devastating effect.

*Not the **tool**, but the **use of the tool** is or is not pedagogical.*
(Vlasta Kokol-Voljc) (in Kutzler, 2005)

Both arguments appear plausible—so what to do?

With every technology, a key question is when to use it and when not to use it. Should we use a car whenever we want to go from A to B? If A and B are a hundred kilometres apart, the answer is "yes." If A and B are only five meters apart, the answer is "no." What makes the "no" turn into a "yes?" Is it a distance? Or is it (also) a purpose? A physical challenge such as a handicap of walking will influence the decision (see the section *Compensate* of this paper).

If A and B are three kilometres apart—should we walk or drive? If the purpose for moving from A to B is shopping, then going by car sounds reasonable. If the purpose is to improve physical fitness, then we should jog—not drive.

This thinking can be applied also to using mathematical tools. As an example, we look at the "black box" solve. When should we use solve?

When we ask our students to solve an equation, there are two possible motivations for that. Either we want the solution—for example, because the solution is needed within a bigger problem—or we want the students to take the steps to the solution so that they improve their (mental) algebra skills. This is exactly as it is with physical movement: When we move, then either we are interested in reaching the destination or we are interested in the moving. The key question in the classroom, therefore, is:

*Are we interested in the **solving** or in the **solution**?*

When we want the *solution*, then we should use technology so that we obtain the solution quickly and can rely on its correctness.

When we want the *solving (process)*, then we should not use technology (except, when necessary, for lower level tasks as explained in the *Compensate* section).

The following is a useful model:

(school) mathematics = mental training + problem-solving training

Educators who desire to ban technology are advocates of mental training. Educators who desire to use technology as much as possible are advocates of problem-solving training.

Nothing is either good or bad—only thinking makes it so.
(William Shakespeare)

School mathematics has both aspects. A CAS forces us to ask questions that we should have asked earlier. The above exposition gives an answer to the question *"What is the purpose of a classroom task?"*

Another question is *"What are indispensable manual skills?"* There is no general answer to this. Using the car metaphor, this is the question as to what distance the students should be able to move without a car. The answer is a matter of definition of what students going to a certain type of school should be able to do without technology. [2]

Assessment is an important pedagogical instrument. Therefore we need to ask *"how to integrate technology into assessment."* A practical answer is easily derived from the "mathematics = mental training + problem-solving training" model. One simply splits the exam in two parts: When assessing mental fitness, no tools are allowed. This includes even simple four-function calculators. When assessing problem-solving capabilities, all tools are solicited. If the split is not manageable within a single exam, one should assess the two "disciplines" at different times.

Here is a parallel with ice skating: Mental training compares with the compulsory exercise, in which the athlete demonstrates a mastery of the basic techniques, and problem solving compares with the voluntary exercise (freestyle), in which the athlete demonstrates the ability to combine the basic techniques into a choreographed presentation. The total score depends on the scores of both the compulsory and the voluntary exercise.

EXPLORE

How do we learn walking, speaking, etc.? We try, we observe, we fail, we analyze, we try again, ...How did mankind discover all the known mathematics? By the same method.

More formally, we can describe the method of mathematical "growth" as follows. Applying known algorithms produces *examples.* From the examples, we *observe* properties that are inductively expressed as a *conjecture.* Proving the conjecture yields a *theorem.* The theorem's constructive parts are *imple-*

[2] A thought-provoking attempt can be found in Herget et a. (2000, 2001), see References.

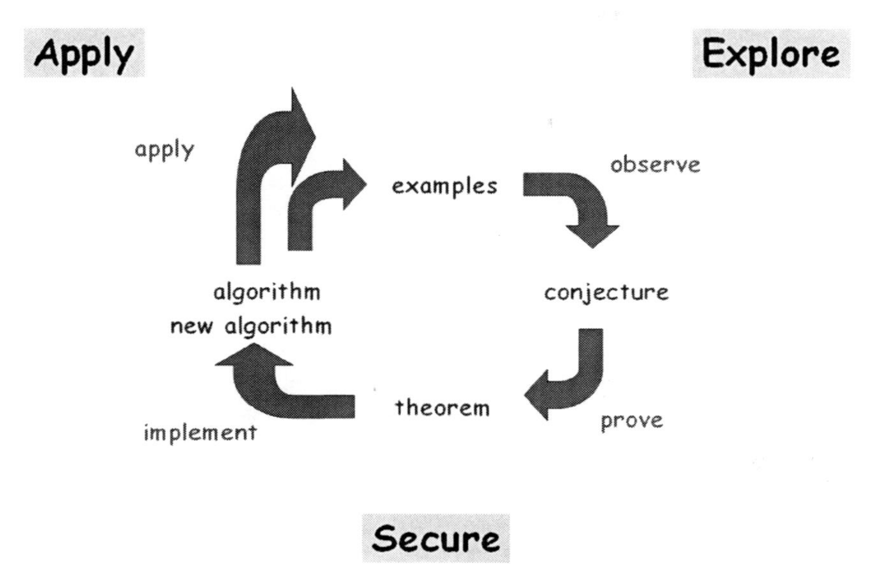

FIGURE 10.

mented in *new algorithms*. Then the old and the new algorithms are *applied* to new data, yielding *new examples* that lead to new observations, new conjectures, and so on.

Figure 10 was proposed by Bruno Buchberger (Heugl, Klinger, & Lechner, 1996). In the spiral we find three phases: *exploring*, *securing*, and *applying*. They can also be denoted as **in**duction, **de**duction, and **pro**duction.

In its beginnings, mathematics was a purely experimental science; i.e. it consisted only of the phases of exploration and application. Later, the Greeks added the phase of securing, thus establishing mathematics as a deductive science. Fairly recently a group around the French mathematician Dieudonné (the group became known as *Bourbaki*) restructured the mathematical knowledge using "definition-theorem-proof-corollary-example..." This Bourbaki system (*Bourbakism*), being developed for inner-mathematical documentation and communication, comprises only the phases of securing and applying and has become characteristic to modern mathematics. But then, Bourbakism gradually lodged itself in teaching and learning. It has become customary to teach mathematics by presenting mathematical knowledge, and then asking the students to learn it and apply what was learned to solve homework and exam problems.

This is unnatural. No mathematician could do mathematical research the way we demand our students to do it.

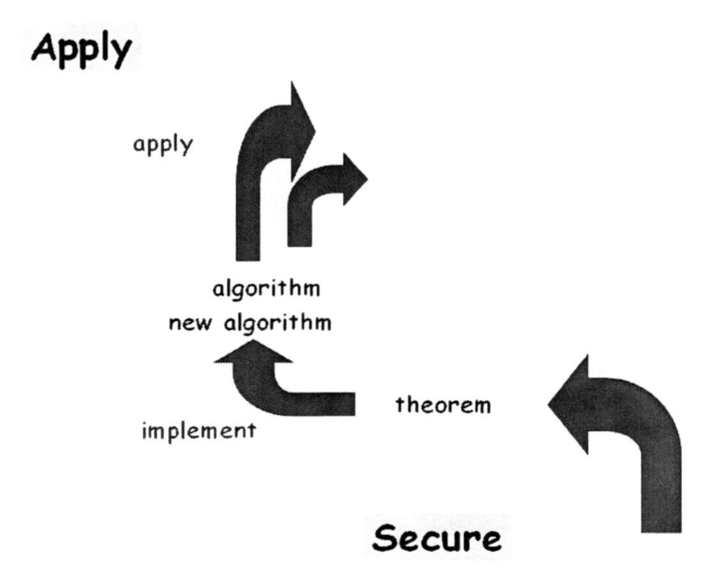

FIGURE 11.

It is probably the Bourbaki style of mathematical documentation that gives the wrong impression that mathematics is not an "experimental science," although it definitely is—to some extent. A good example is Andrew Wiles' proof of Fermat's Last Theorem. Wiles worked for about seven years on this proof, and obviously he spent most of the time in the phase of exploration. The Bourbaki-style summary of his work is a 109-page article.

A student has to "locally" build his individual little "house of mathematics," while a scientist does the same "globally" by searching for mathematical knowledge that is new for mankind. Phases of exploration should complement traditional teaching methods—not substitute for them. This is not a plea for returning to pre-deductive Egyptian mathematics, but a plea for mathematics teaching and learning going through all three phases of the spiral.

Within today's mathematics curricula there is no place for exploration. With paper and pencil students can generate only a very small number of hand produced examples for the purpose of observing and discovering, and a hefty portion of these probably would be faulty due to calculation and other errors. Nothing can be observed from only a few, partly wrong examples!

Say, we want to teach our students that in every triangle the three altitudes intersect in one point. We ask them to draw five triangles, and then con-

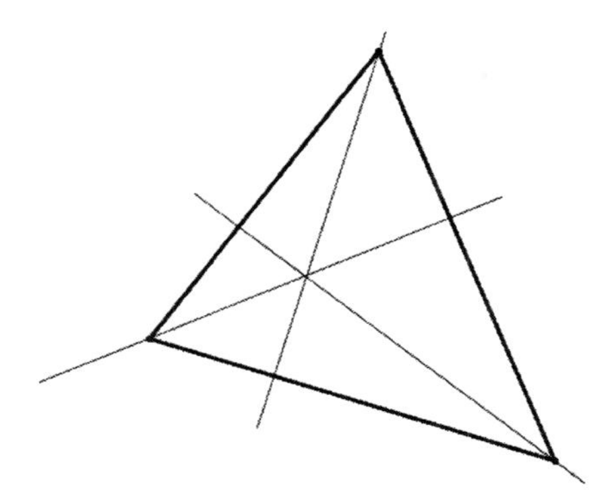

FIGURE 12.

struct the three altitudes in each. Most of our students—being lousy drafts(wo)men—will find that in three or four of their triangles the three altitudes do NOT intersect in one point!

Technology enables students to experiment within almost all topics treated in mathematics teaching. The plea for allowing students to find what they are supposed to learn is not new.

Help me to do it by myself.
(Maria Montessori)

We should not teach students something that they could discover themselves.
(Hans Freudenthal)

With technology we can implement a new teaching culture that could be called *guided explorations*, in which the teacher observes the students in their experiments and feeds them with useful hints along their "journey" in order to help them reach the goal and make the intended discoveries.

Give a person a fish and you feed them for a day.
Teach a person to fish and you feed them for a lifetime.
(Confucius)

Give a student some mathematics and you feed them for the next exam.
Teach a student to fish for mathematics and you feed them for a lifetime.
(Confucius, adapted)

CASANOVA OR DON JUAN?

Giacomo Girolamo *Casanova* was an Italian adventurer and writer who lived 1752–1798. *Don Juan* is a legend, used as hero in opera, play, and fiction. Both are famous womanizers—though there is a significant difference: Casanova wanted pleasure for the women, Don Juan wanted pleasure for himself.

We use this difference for a classification of teachers, notably

- Casanova-type teachers and
- Don Juan-type teachers.

The student is in the centre of all teaching.

A Casanova-type teacher meets the student where he is and guides him through the topic of teaching as far as this student can go. The student comes first, mathematics comes second. Every (group of) student(s) is a new challenge and the teaching is always different.

For a Don Juan-type teacher, mathematics comes first and the student comes second. Their teaching is always more or less the same, notably a "sink or swim" style.

A Casanova-type teacher *teaches students.*

A Don Juan-type teacher *teaches mathematics.*

As said before, students are the original goal of all teaching.

Therefore, we should teach students.

Therefore, we should be Casanovas.

After all that you have read in this paper, you will understand my plea that …

*… we should be **CAS**anovas.*

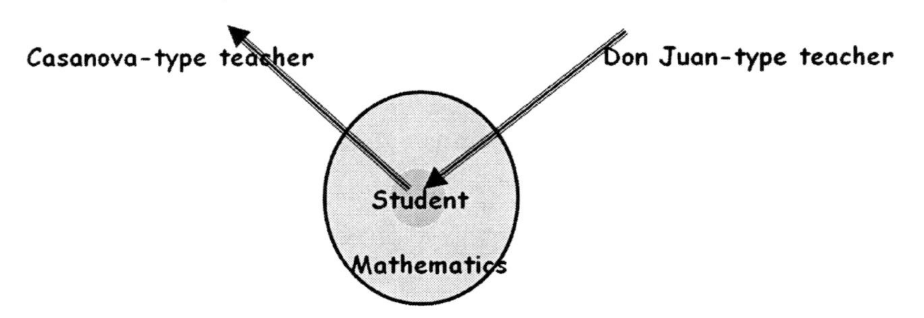

FIGURE 13.

REFERENCES

Aspetsberger, K., & Funk, G. (1984). Experiments with muMATH in Austrian high schools. *ACM SIGSAM Bulletin, 18-19*(4-1), 4–7.

Buchberger, B. (1989). *Why should students learn integration rules?* RISC-Linz Technical Report no. 89-7.0, University of Linz: Austria.

Daily Montessori. (2009). *Biography of Maria Montessori.* Retrieved February 18, 2009, from http://www.dailymontessori.com/dr-maria-montessori/.

Herget, W., Heugl, H., Kutzler, B., & Lehmann, E. (2000, 2001). Indispensable manual calculation skills in a CAS environment. *Ohio Journal of School Mathematics, Autumn 2000*(42), 13–20; also: *Micromath, 16*(3), 8–12; also: *Mathematics in School, 30*(2), 2–6.

Heugl, H., Klinger, W., & Lechner, J. (1996). Mathematikunterricht mit Computeralgebra-Systemen (Ein didaktisches Lehrerbuch mit Erfahrungen aus dem österreichischen DERIVE-Projekt). Bonn, Germany: Addison-Wesley.

Kutzler, B. (1995, 1996). *Improving mathematics teaching with DERIVE.* Bromley: Chartwell-Bratt.

Kutzler, B. (2003). Computers and calculators as pedagogical tools for teaching mathematics. In J. T. Fey, A. Cuoco, C. Kieran, L. Mullin, & R. M. Zbiek (Eds.), *Computer Algebra Systems in Secondary School Mathematics Education* (pp. 53–71) Reston, VA: NCTM.

Kutzler, B. (2005). *Zitate von Mathematikern und Zitate über Mathematik.* Linz, Austria: Author.

Kutzler, B. (2008). *Technology and the Yin and Yang of teaching and learning mathematics.* Linz, Austria: Author.

CHAPTER 7

CAS AND THE FUTURE OF THE ALGEBRA CURRICULUM

Kaye Stacey

In this contribution, I was asked to explain the focus of my work on the use of CAS in school mathematics and how this relates to changes in the school mathematics in my country over the last decade, and the influence of research on this curriculum change; the implications of these changes for the teaching of reasoning and proof; the implications for pre-service and in-service teacher education; and to present a vision for the next ten years. Since my research work on CAS has been tightly interwoven with major curriculum changes I will address both of these in the first part of the article, ending with some brief notes on the implications of the changes for pre-service and in-service teacher education. In the second part of the paper, I will discuss some of the ways in which using CAS affects teaching and learning mathematics, and I will end with a look ten years ahead.

CAS IN SCHOOLS IN VICTORIA

Research on various forms of mathematically able software has been continuing at the University of Melbourne with a team of researchers (including Stacey, McCrae, Asp, Dowsey, Tynan, Kendal, Ball, Flynn, and Pierce) since the early 1990s. The term "mathematically able software" refers to

Future Curricular Trends in School Algebra and Geometry: Proceedings of a Conference. pages 93–107

computers and calculators equipped with the capability of carrying out standard mathematical processes, including graphing, spreadsheet and list functionality, geometric transformations, statistics, and more recently symbolic algebra manipulation. This article is specifically focussed on CAS (computer algebra systems) which includes numerical, graphical and symbolic manipulation. It is symbolic manipulation, the ability to work with algebraic letters and not just numbers, that is the most challenging from a curriculum perspective. Since our work has been closely related to changes in the school situation in Victoria, the state of Australia where we work, I will report the foci of our work and the school situation together. In this state, the strongest determinant of school curriculum, and the one most relevant to CAS, is the end-of-school examination system. These examinations are closely prescribed by the state, and are very high stakes for teachers and students, since competitive access to university courses depends on the results.

Graphics calculators (without a CAS facility) were first permitted in the state Year 12 examinations in the mid 1990s. There was a serious concern with the fairness of allowing students to use such an expensive machine, so at first graphics calculators were permitted but not really required. We termed the questions that made up such an examination "graphics calculator neutral." In a calculator neutral question, having access to a graphics calculator is not an advantage, although it may provide alternative methods of solution and of checking the answer. Since 1998 (when the price of the calculators dropped somewhat) the examinations for the major mathematics subject ("Mathematical Methods") have been "graphics calculator active," so that some questions are included where having a graphics calculator makes the question very much easier than a by-hand solution and a student without the calculator would be disadvantaged. Use of graphics calculators is now very well accepted by most teachers, and has proved very successful. Having ready access to accurate graphs with manipulable graphing windows has strengthened the graphical approach to the functions and calculus strands that dominate "Mathematical Methods" and many teachers find this increases students' understanding. On the other hand, after a decade, there are still a few teachers who are not confident users!

As part of the preparation for the introduction of graphics calculators into Years 11 and 12 mathematics, McCrae and Flynn (2001) analysed the traditional examination questions to see the effect of a graphics calculator with and without symbolic algebra manipulation. They found that the solutions of a minority of exam questions would be affected by graphics calculator use but around 60% of exam questions would be affected by having symbolic algebra and many of these would be trivialised. As a consequence, it was clear that the introduction of full CAS was a much more substantial curriculum step than the introduction of graphics calculators. The main obstacles to introducing (non-CAS) graphics calculators had been equity (the need to purchase expensive equipment) and teacher training, but the main obstacle for introducing CAS calculators was to be the curriculum issue.

INTRODUCING CAS TO SCHOOL MATHEMATICS

After some introductory experiments examining the issues in using CAS for teaching calculus in Year 11 (see, for example, Kendal, 2001), arrangements were made for the piloting of a course, "Mathematical Methods CAS," where students would have open access to CAS (on one of three brands of calculator) during lessons, at home, and in all the state examinations (Leigh-Lancaster, 2003; Victorian Curriculum and Assessment Authority, 2004). Teachers could choose the extent to which students used the calculators in class on any given day, but they would be permitted in the external assessment. The CAS-CAT research project, conducted by the University of Melbourne team, was established with funding from the Australian Research Council and the Victorian Board of Studies, to provide academic support to the curriculum development and the pilot teachers, and to conduct research into the outcomes of the project. It is highly unusual that curriculum change is accompanied by educational research in this way. The CAS-CAT website (Melbourne Graduate School of Education, Science, and Mathematics Education, 2006) provides links to the many papers that resulted. The new subject proved a successful alternative to the normal Mathematical Methods subject that permitted only a graphics calculator, and a growing number of schools began to offer it. At the time of writing, it is envisaged that soon the graphics-calculator-only subject will be phased out, and all students in this subject will be able to use CAS. As a consequence of the planned change, there is now growing interest by teachers in using CAS at Years 9 and 10, because school policies will encourage students to purchase their calculator at that age. After several years of allowing CAS in all assessment, a "technology-free" component of the examination was introduced to encourage students to pay attention to by-hand skills, along the line of the AP Calculus subject. It is still too early to comment on the effect of this change.

In this section, I will give an extremely brief summary of the findings of the CAS-CAT project. The teachers reacted positively to the new technology, as would be expected from volunteers. At the beginning, some found it difficult to learn the new technical skills required, and to teach them to their students, but their confidence grew over the life of the project, and they began to see more ways of enhancing their teaching with CAS. The students were also nearly all positive. We investigated whether students' by-hand skills were less than students who did not have CAS. There was no apparent difference. This was due to the importance that teachers gave to developing adequate by-hand skills; for many teachers in these early years of using CAS, "understanding mathematics" meant being able to perform routines by-hand. All teachers noted that the support of CAS enabled a few students in their class with weaker than expected algebraic skills to undertake Mathematical Methods CAS, and they saw this increased access to the subject as a desirable thing. Daily experience showed that the symbolic aspects of CAS are complicated to use, but all teachers found it manageable. They noted that students need good "algebraic expectation" (a concept paralleling the "number sense" that is required to sensibly use a four-function calculator), and a short quiz was developed to assist teachers in communicating the importance of this ability to students (Ball, Stacey & Pierce, 2003; Melbourne Graduate School of Education, Science, and Mathematics Education, 2006; Pierce & Stacey, 2002). Although it was important not to increase expectations too much, it was found that the curriculum could be expanded somewhat as teaching time could be shifted from practising routine skills. For example, the examination syllabus placed fewer restrictions on types of functions that could be used in the examination, and a few topics were added. One successful addition was continuous probability distributions, which came late in the year and brought together the work on calculus with the work on probability. Transition matrices, for which by-hand calculations could be too time-consuming, were also added successfully, again linking into the existing probability work. Using CAS also required increased emphasis on some topics. For example, absolute value played a minor notational role previously, but its very common occurrence in solutions presented by CAS gave it more prominence and so functions incorporating absolute values were studied more than previously.

Setting good examination questions was a challenging task, especially as there is a tendency to believe that any question that simply asks for a routine procedure and can be done by CAS should be discarded. With this policy, it is easy to make assessment with CAS far too difficult for most students, since the reality is that many students only score marks on the routine items. Flynn (2001) studied how CAS could be used to expand the "bandwidth" of mathematical competence that an examination tested, from routine procedures to questions requiring insight and problem solving skills.

At the beginning of the project, the researchers and curriculum developers set out their goals for using CAS (Stacey, Asp & McCrae, 2000). In broad summary, the researchers wanted students to have more mathematical power. They wanted students' mathematical performance to be amplified by having CAS, so that they became better users of mathematics, experienced more real problem-solving and worked in an environment where there was a reduced dominance of routine procedures in the classroom. Both researchers and teachers wanted better learning, especially through the use of multiple representations, and by capitalizing on the positive learning strategies associated with collaborative learning that technology use often encourages in classrooms. Teachers had "better learning" as their overarching goal. They also wanted access to mathematics for more learners, because they saw that CAS could be used to compensate for poor algebraic skills. This was something that the researchers had not predicted. Where the researchers wanted amplification of the curriculum, the teachers valued compensation and increased access more.

Researchers and teachers also differed in the impact that they saw CAS use having on the curriculum content. The researchers and examination committee were concerned with how the power of CAS would affect how mathematical problems were solved, and consequently what topics should be taught in the new subject. In general, the teachers took a long time to understand the extent of the mathematical power of CAS, and throughout the project were regularly surprised to see its capabilities. For example, some were very surprised when they inputted $d(f(x).g(x))/dx$ and saw that CAS "knows" the product rule of differentiation, having previously treated CAS more like a table of derivatives. Only towards the end of the project did they slowly begin to re-examine current curriculum values in the light of the new technology, and to ask themselves why each topic was included in the syllabus and why students should learn it. To some extent, the recent decision to have a technology-free component of the examination has reduced the need for teachers to re-evaluate the reason for the inclusion of each topic in the curriculum. Having to prepare for the "technology-free" examination provides sufficient reason for learning any routine procedure, and the time that it takes to polish skills means that there is little possibility to seriously consider further amplifying the curriculum. This may change with time.

In summary, over ten years, it has become accepted that mathematically able software including CAS should be a regular tool in the mathematics classroom. Teachers have generally been positive. They have been concerned that students develop adequate by-hand skills (which many teachers still see as defining "understanding"), and have created classroom practices where both by-hand and technology use co-exist. CAS has also been accepted as a regular tool in mathematics assessment, including the crucial end-of school, university entrance examinations.

PRE-SERVICE AND IN-SERVICE EDUCATION

The implications for the changes on pre-service and in-service education have been extensive. University mathematics discipline courses vary considerably in the use that is made of technology. For over a decade, students have arrived at university already possessing a graphics calculator, which in some places they continue to use. Many universities prefer to use computer technology instead or as well (especially for those learning mathematics with an engineering orientation), and a few courses in a few universities turn their back on all material aids to doing mathematics except pen and paper. It is be hoped that over the next ten years, more consistent use will be made of the technology skills that students bring with them from school in their tertiary courses. Within teacher preparation courses, students routinely learn to teach with graphics calculators, and increasingly with CAS calculators. Older pre-service teachers who come to teacher education without personal experience of graphics or CAS calculators may need more assistance than is offered.

The in-service education requirements for the changes outlined above have been very substantial. The state department of education and the Catholic education system have provided a basic level of training for all mathematics teachers, first for graphics calculators and more recently for CAS calculators. Calculator companies, the mathematics professional association, and private consultants have been actively sought by schools to provide more training. My own university has provided extensive in-service education through special subjects in the Master of Education degree and, for a greater number of teachers, through non-award courses (Melbourne Graduate School of Education, Science, and Mathematics Education, 2006). Even after a decade of graphics calculators and five years of CAS, these courses are still well-subscribed. There is continuing demand from teachers of Years 11 and 12, but also a growing demand from teachers of Years 9 and 10. Students in Years 9 and 10 are purchasing CAS calculators in preparation for Years 11 and 12. Teachers therefore want to know how to use them to enhance students' learning of basic algebra. Some of the most promising uses are as follows. Learning to solve equations in the home screen (e.g. learning the "do the same to both sides" method is well scaffolded by CAS because it can deal with the manipulation whilst the student learns the strategy. Students' early learning of the strategies of equation solving is often clouded when they make manipulation errors. The ability of CAS to work with exact values (for square roots and trigonometry, for example) assists in giving meaning and importance to non-numerical calculations. Solving simultaneous equations is another rich topic because it involves multiple representations, and CAS can scaffold the learning of the different strategies for by-hand solution. Our aim in professional development sessions is to improve mathematics knowledge and teaching skills, in sessions focussed on calculator use. It is now our belief that teachers' skills and interest to become a competent user of technology is as much an equity issue for students as is providing the machines at reasonable cost.

In the following sections, I will give examples to highlight some implications of the adoption of CAS on the school mathematics curriculum (and particularly on reasoning and proof), drawn from the experiences in the research projects described above.

CAS GIVES STUDENTS A MORE POWERFUL VIEW OF FUNCTIONS

From our first experiments with function graphers in the early 1990s, it has been our consistent observation that the necessity to communicate with technology by using function notation has had a good impact on students' understanding of function. These observations supported our use of function notation much earlier than is traditional in "Graphic Algebra" (Asp, Dowsey, Tynan & Stacey, 1998). Using a label to identify a function appears to assist in the transition of function from "process" to "product. " Instead of accepting y = 3*x + 1 or y1 = 3*x + 1 for example, one of the early function graphers that we worked with required f(x) = 3*x + 1, g(x) = cos(x) etc. This seemed to have positive benefits that outweighed the difficulties of dealing with the new notation. In our first experiments with CAS, students first began to define functions to avoid having to enter long expressions on multiple occasions with the awkward one-line entry syntax. Defining the function, and later calling it up by name, seemed to assist the process of seeing the functions in a question as objects that could be manipulated.

These observations are illustrated with the question in Figure 1, which has significant health importance to teenagers. Solving this problem shows the power of the multiple representations available in CAS. Students can readily make tables of the BAC readings, to see that the BAC is over the legal driving limit from 15 minutes to 3 hours and that the maximum BAC reached is about 0.10 after an hour. The table can be readily linked with a graph, as shown in Figure 2. Alternatively, the symbolic features can assist students to find the maximum value by calculus and solve the resulting equations.

A model for the BAC (blood alcohol concentration) after t hours of a man who drinks V gm of alcohol is $c(t) = V(e^{-0.85t} - e^{-t})$.

A young man drinks 170 gm of alcohol almost at once. (This is a lot of alcohol – the gram measurement refers to the alcohol content, not the volume of the drink).

Explain why the BAC function is likely to be the difference between two exponential functions.

What is the maximum BAC reached?

A safe BAC for driving is 0.05. When is it not safe for the young man to drive?

Explore what happens to the BAC if the young man drinks the same quantity of alcohol in two equal drinks an hour apart, 3 equal drinks an hour apart, etc.

FIGURE 1. A good question to illustrate uses of CAS

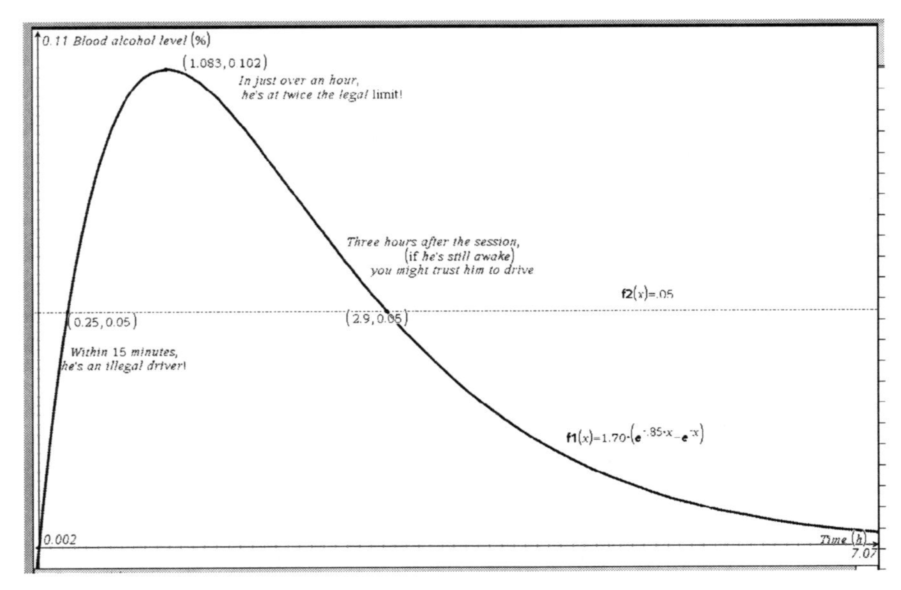

FIGURE 2. Graph of BAC as a function of time.

A key feature of symbolic manipulation is that it can make function a highly manipulable object. Computer and calculator graphing has been successful in our schools because it has made a graph a manipulable object. In the by-hand environment, a graph was static on a page, but now we have scale and window changes that enable a graph to answer multiple questions, often not foreseen when the original graph was drawn. Symbolic algebra also needs to give us easy power to manipulate mathematical objects to be a really successful tool in school.

Exploring what happens to BAC if the alcohol is consumed in several drinks, rather than at once, shows the sorts of manipulations of functions that are possible. Drinking half the alcohol at the first drink involves the function $0.5c(t)$, instead of $c(t)$. Delaying the second drink for an hour requires the transformation of $c(t)$ to $c(t-1)$. By communicating with the CAS in terms of transformations of $c(t)$, a graph for the BAC of multiple drinks one hour apart can readily be obtained, as shown in Figure 3 for 2, 3, 5 and 6 equal drinks. The expression for the function $c_2(t)$, the BAC when the alcohol is consumed in two equal drinks an hour apart, is given below.

$$\begin{cases} c_2(t) = 0.5 * c(t) \text{ for } t < 1 \\ c_2(t) = 0.5 * c(t) + 0.5 * c(t-1) \text{ for } t > 1. \end{cases}$$

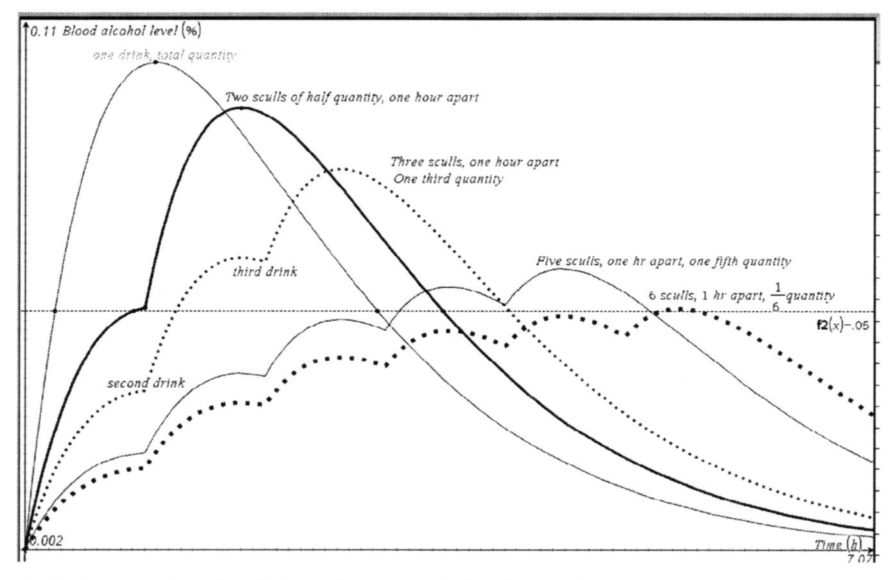

FIGURE 3. Graph of BAC for equal drinks an hour apart.

As with many real problems, this problem requires hybrid functions that have been awkward to treat without technology. Issues of domain of definition have previously been dealt with ritualistically in standard mathematics courses (find the points where a function is undefined because there would be division by zero, and exclude them from the domain). However, with real world problems, domain can become critical to the formulation and solving processes. CAS and graphing software provide an improved way of dealing with these important ideas by making work with hybrid functions in real world contexts accessible.

POTENTIAL CONFLICTS BETWEEN CAS AND MATHEMATICAL REASONING

In this section, two of the ways in which CAS use can conflict with the development of mathematical reasoning are shown. The first is the way in which technology supports exploration of mathematical ideas. This is a very considerable strength of CAS, but teachers must not let exploration remain at the inductive level. The second example shows how algebraic manipulation, which is arguably the principal mode of reasoning in high school mathematics courses, is much reduced when students work with CAS, but is replaced by a much greater emphasis on formulating problems mathematically and interpreting results. The resolution of this potential conflict of using CAS with developing mathematical reasoning lies in curricula giving more explicit attention to the reasons for including each mathematical technique.

A popular use of CAS in many of the classrooms that we observe is to use it to explore mathematical situations and to find patterns. For example, a teacher beginning work on expanding linear factors may have students expand the quadratic expressions in Figure 4 using CAS, and then look for a pattern. The teacher hopes that students will note that the coefficient of x in the product is the sum of the numbers in the factors, and the constant term is their product.

It appears to me that an approach like this is popular with teachers because they believe that it gives students some ownership of the result if they discover it themselves. However, it is often the case, that teachers leave such results at the level of inductive reasoning, and do not demonstrate, or have students independently engage with, the deductive reasoning that characterizes mathematical proof. This strong support for gathering data and pattern spotting that CAS provides is one feature that may adversely affect the teaching of deductive reasoning in some classrooms. An analogous situation has been debated in the research literature on dynamic geometry: students can be convinced visually by dragging a geometric figure that a certain property is always true, and this may make a deductive proof an apparently unnecessary after-thought. As with dynamic geometry the use of CAS to provide data for pattern spotting need not necessarily lead to neglect of deduction, but including deduction requires commitment on the part of the teacher.

Using CAS significantly affects the most common type of reasoning in high school algebra: proof by algebraic manipulation. Basic algebraic manipulation is required to use CAS, since users need good "algebraic insight" (Pierce & Stacey, 2002) to monitor their work, but the necessity for complicated manipulation is removed. However, whilst it removes the need for the algebraic manipulation, using CAS can require very strong skills of formulation, which could be argued to require deeper algebraic skills than manipulation. Consider, for example, the problem of finding whether the graphs of all cubic polynomials have symmetry. Examining a few special cases (see Figure 5) quickly shows that if cubic graphs have symmetry, it must be rotational symmetry. Proving this by hand requires many steps of algebra.

$(x+3)(x+6) = x^2+9x+18$ $\quad (x+2)(x+7) = x^2+9x+14$ $\quad (x+1)(x+4) = x^2+5x+4$

FIGURE 4. CAS can provide data for students to find patterns.

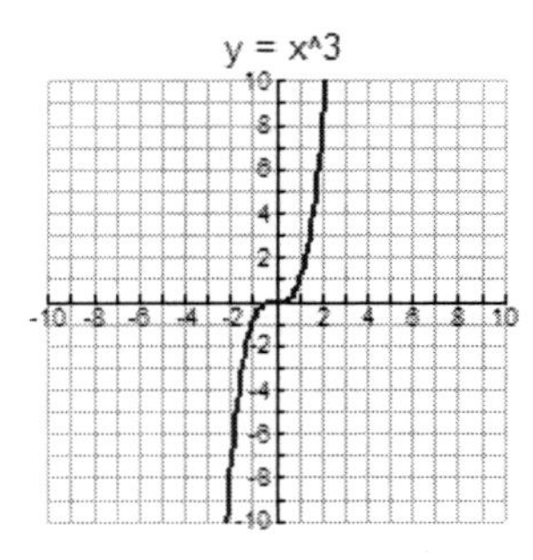

Are the graphs of all cubic polynomials symmetric?

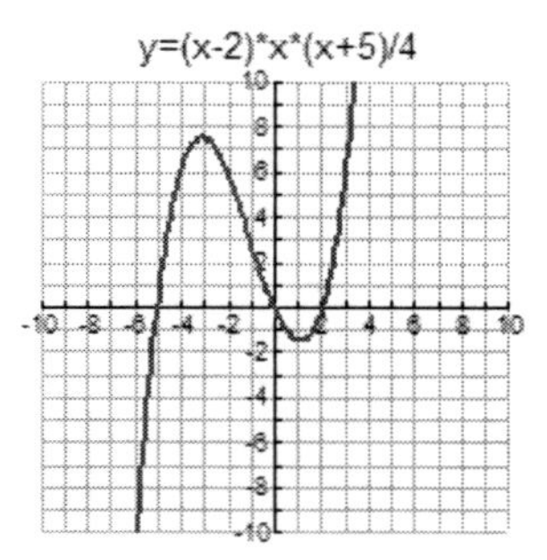

FIGURE 5. Examples show that cubic graphs can only have rotational symmetry

Proving this result by CAS has an unfamiliar feel to it. The key step is to formulate the required question mathematically, write in an appropriate CAS syntax, and then to interpret the result that CAS gives. If there is rotational symmetry around the point $(s, f(s))$, then for all values of h, equation (2) in Figure 6 will hold. Solving this expression for s gives the solution as shown in Figure 6 (3). The key observations are that this solution for s does

$$\textit{Define}\ \ f(x) = ax^3 + bx^2 + cx + d \qquad (1)$$

$$\textit{Solve}\ \ f(s+h) - f(s) = f(s) - s(s-h)\ \textit{for}\ s \qquad (2)$$

$$\text{Solution: } s = \frac{-b}{3a} \qquad (3)$$

FIGURE 6. Using CAS emphasizes formulating problems mathematically.

not involve h, and that it therefore identifies the centre of symmetry. Readers will recognize this as the point of inflexion of the cubic graph. Geometric thinking shows that this is the only candidate for the centre of symmetry because if a different centre of symmetry existed, both the point of inflexion and its (distinct) image would be points of inflexion yet a cubic graph has only one. This problem and many other examples show that when using CAS the emphasis shifts from transforming mathematical expressions to formulating problems mathematically and interpreting the results.

As has been evident since CAS first came onto the horizon of school mathematics, using CAS presents very fundamental challenges to the mathematics curriculum. The examples above illustrate two of the many ways in which this may happen—the first that teachers may use the example-generating facility to bypass mathematical reasoning and the second that using CAS shifts the intellectual demand from many routine steps to the difficult formulating and interpreting steps. There are many other examples of associated challenges; for example numerical and graphical facilities can lead students and teachers to turn away from the symbolic methods that are beneficial in the long term.

The French school (Artigue, 2002) distinguishes two purposes for mathematical techniques in the curriculum. Some curriculum topics have a pragmatic purpose: to find a solution to one equation for example. It does not matter whether this is solved numerically, graphically or symbolically; what matters is finding the solution. For nearly all of school mathematics, pressing a button to find the numerical value of sine of an angle is sufficient—students do not need to know how it is calculated by the machine. Other curriculum topics have epistemic value. They help us connect mathematical ideas; they demonstrate reasons why an idea works; they build a conceptual foundation for future work. I think that almost all curriculum developers would see that finding the rule for expanding the expressions in Figure 4 simply by pattern guessing, without reference to the distributive law, would be missing out on the epistemic value of this topic. In our analysis of the impact of CAS on school curriculum, we have found it useful to add a third source of curriculum value to the French analysis: pedagogical value (see Figure 7). In designing a curriculum, there may well be topics

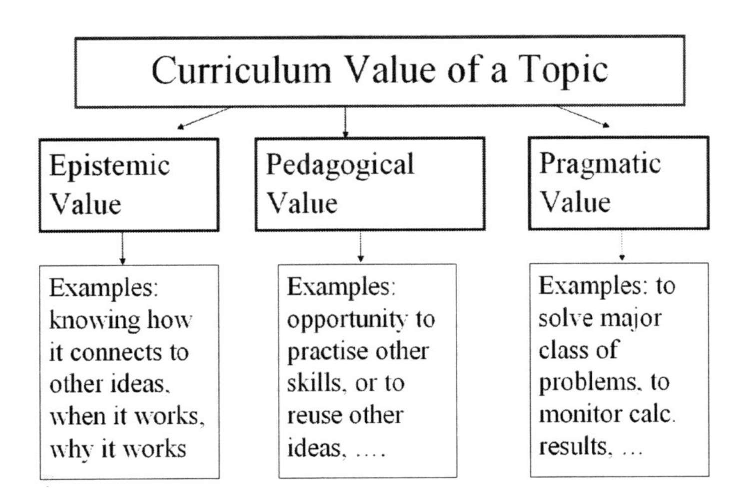

FIGURE 7. Three sources of curriculum value.

that are not justified primarily by either their pragmatic or their epistemic value, but instead by the way in which they provide an avenue for the development and consolidation of learning of other important topics. Some by-hand symbolic skills of moderate complexity may well fit into this category in particular curricula. Curriculum developers need to consider carefully the value of the topics and techniques that they include in mathematics for the future. In the pre-electronic age, symbolic manipulations were a pragmatic aid to numerical calculation, even at the school level, in addition to their role in doing algebra pragmatically and epistemically. In the numerical calculation age, algebraic simplifications were for doing algebra. Now they shift principally into the epistemic realm, and carefully into the pedagogical realm, demanding much closer scrutiny.

A WISH LIST FOR THE NEXT DECADE

Over the next decade, I expect that more school systems will incorporate mathematically able software into mathematics teaching. Substantial issues of equity can be solved at the class and school level, by active assistance in equipping individual students, and to some extent between different parts of one country. We should also work towards gradually decreasing gaps between countries.

Within ten years, I hope for software that is easier to use and looks more like by-hand mathematics. Mathematically able software, especially symbolic manipulation, is complicated software that is difficult to learn to use. Many of these complications are inherent within the mathematics. For example, a human problem solver may assume from context that a variable has to be positive or that a solution has to be real, and so can write a simple expression for a solution for an equation. However, unless the constraints are made explicit, a CAS solution must consider all possibilities and so the solutions may be very complicated. Despite these inherent constraints, further development of software and hardware may provide us with modules that are well adapted to the stages of different learners, with easier entry points.

At the same time, mathematically able software could be incorporated into a complete presentation tool for teaching, with classroom connectivity, facilitating sharing between students and sending real-time information to the teacher. This would open up new possibilities for mathematical explorations in classrooms that provoke and support inquiry, conjecturing and convincing. The present-day focus on learning how to use the technology, which seems inevitable in so much of today's professional development for teachers, should be able to be reduced with a consequent refocussing on mathematical content and teaching development.

Within ten years, I hope that we will have created curricula that use mathematically able software to amplify students' mathematical abilities as well as meeting the teachers' goals of better learning and increased access to mathematics. My experience with graphics calculators is that they have amplified what students can do. Many students can solve problems by creating tables of values and drawing graphs, and using them to find numerical solutions to problems. Over the decade or so of use of graphics calculators in our schools, this has caused a "numerical turn" in students' thinking: they go first to a graph or table of values, rather than to a symbolic algebra technique. I hope that within ten years, with the support of technology that speaks the language of symbolic algebra, we have amplified what students can do symbolically and given them correspondingly increased power to tackle problems amenable to a generalised solution, within and outside mathematics. All of this depends on seeing formulating problems in mathematical (usually in algebraic) terms as a more important process for mathematics learning than carrying out routine procedures. In ten years, I hope that the balance will have significantly shifted, as more teachers consider the importance of mathematical modelling and mathematical reasoning as outcomes of school mathematics. To do this, we also need a better articulated vision of what it means to understand mathematics, other than to carry out routine procedures accurately and quickly by hand. We need to work together on a careful reassessment of all curriculum topics from epistemic, pragmatic and pedagogical viewpoints.

Through processes such as this, I hope that we come closer to realising the vision that was put forward by Dan Kennedy (1995) when he likened mathematics education to climbing a wonderful tree. In the branches of this tree there is much to explore and there are great views of important places to be seen. He likened traditional approaches to mathematics teaching as having students all attempt a long and arduous climb up the trunk of the tree, with many falling off, some blocking the progress of others, and only a few reaching the branches, exhausted, near the end of their allotted time in the tree. Instead, Kennedy proposed that we find ladders to get students into the branches more quickly and more reliably, so more of them can begin to see mathematics as something powerful for their lives. Mathematically able software is one of the ladders we can use, and I hope that in ten years, we know better how to use it in this way.

REFERENCES

Artigue, M. (2002). Learning mathematics in a CAS environment: The genesis of a reflection about instrumentalisation and the dialectics between technical and conceptual work. *International Journal of Computers for Mathematical Learning, 7*, 245–274.

Asp, G., Dowsey, J., Stacey, K., & Tynan, D. (1998). Graphic Algebra: Explorations with a Graphing Calculator. Berkeley: Key Curriculum Press.

Ball, L., Stacey, K., & Pierce, R. (2003). Recognising equivalent algebraic expressions: An important component of algebraic expectation for working with CAS. In N.A. Pateman, B.J. Dougherty and J.T. Zilliox (Eds.). *Proceedings of the 27th Conference of the International Group for the Psychology of Mathematics Education, 4,* 15 –22. Honolulu, HI: College of Education, University of Hawaii.

Flynn, P. (2001). Examining mathematics with CAS: Issues and possibilities. In C. Vale, J. Horwood, & J. Roumeliotis (Eds.). *2001 A Mathematical Odyssey. Proceedings of the 38th Annual Conference of the MAV* (pp. 285–298). Melbourne: Mathematical Association of Victoria.

Kendal, Margaret (2001). Teaching and learning introductory differential calculus with a computer algebra system. PhD thesis, Department of Science and Mathematics Education, The University of Melbourne. Retrieved June 16, 2008, from http://eprints.infodiv.unimelb.edu.au/archive/00001525/.

Kennedy, D. (1995). Climbing around the tree of mathematics. *Mathematics Teacher,* 88(6), 460–465.

Leigh-Lancaster, D. (2003). The Victorian Curriculum and Assessment Authority Mathematical Methods Computer Algebra pilot study and examinations. Paper presented at 2003 symposium of *Computer Algebra in Mathematics Education.* Retrieved June 16, 2008, from http://www.lkl.ac.uk/research/came/events/reims/1-Reaction-LeighLancaster.doc.

McCrae, B. & Flynn, P. (2001). Assessing the impact of CAS calculators on mathematics examinations. In B. Lee (Ed.), Mathematics Shaping Australia. *Proceedings of the 18th Biennial Conference of the Australian Association of Mathematics Teachers.* CD-ROM. Adelaide: AAMT

Melbourne Graduate School of Education (Science and Mathematics Education), The University of Melbourne, Australia. (2006). *CAS-CAT project.* Retrieved August 27, 2008, from http://extranet.edfac.unimelb.edu.au/DSME/CAS-CAT/.

Pierce, R., & Stacey, K. (2002). Algebraic insight: The algebra needed to use Computer Algebra Systems. *Mathematics Teacher, 95*(8), 622–627.

Stacey, K., Asp, G., & McCrae, B. (2000). Goals for a CAS-active senior mathematics curriculum. In M. O. J. Thomas (Ed.). *Proceedings of TIME 2000 An International Conference on Technology in Mathematics Education* (pp. 244–252). Auckland: University of Auckland & Auckland University of Technology.

Victorian Curriculum and Assessment Authority. (2008). *Mathematical methods CAS index.* Retrieved June 16, 2008, from http://www.vca.vic.edu/vce/studies/mathematics/cas/casindex.html.

CHAPTER 8

ALGEBRA IN THE AGE OF CAS: IMPLICATIONS FOR THE HIGH SCHOOL CURRICULUM

Examples from The CME Project

Al Cuoco

INTRODUCTION

This paper builds on my talk at the CSMC meeting and discusses how a new high school curriculum, *The CME Project* (Education Development Center, 2009), uses a computer algebra system (CAS) to help students understand some central concepts in advanced algebra.

The CME Project is an NSF-funded high school curriculum, published by Pearson. It follows the traditional American course structure of Algebra 1, Geometry, Algebra 2, and Precalculus. More details about the program are at www.edc.org/cmeproject. The features of *The CME Project* that are relevant to this paper are:

1. The program is organized around mathematical habits of mind.
2. It uses the TI-Nspire™ environments throughout all four courses and makes essential use of a CAS in the last two years.

Future Curricular Trends in School Algebra and Geometry: Proceedings of a Conference. pages 109–128

Both of these features have implications for how algebra is developed in the program.

The structure for the paper is as follows:

- In the next section, I discuss the habits of mind approach to curriculum development and how it evolved.
- After that, I describe the *The CME Project*'s organizing principles for CAS use.
- Then I give examples of each of these uses, all centered around algebraic investigations into an important class of polynomials.
- Finally, in the last section, I make some conclusions.

THE HABITS OF MIND APPROACH

Almost 40 years ago, early in my high school teaching career, I came to understand that the real utility of mathematics for many students comes from the kind of thinking that is indigenous to the discipline. I have put it this way (Cuoco, 1998):

I didn't always feel this way about mathematics. When I started teaching high school, I thought that mathematics was an ever-growing body of knowledge. Algebra was about equations, geometry was about space, arithmetic was about numbers; every branch of mathematics was about some particular mathematical objects. Gradually, I began to realize that what my students (some of them, anyway) were really taking away from my classes was a *style of work* that manifested itself between the lines in our discussions about triangles and polynomials and sample spaces. I began to see my discipline not only as a collection of results and conjectures, but also as a collection of *habits of mind.*

This focus on mathematical ways of thinking has been the emphasis in my classes and curriculum writing ever since, and I'm now convinced that, more than any specific result or skill, more than the Pythagorean theorem or the fundamental theorem of algebra, these mathematical habits of mind are the most important things students can take away from their mathematics education. For *all* students, whether they eventually build houses, run businesses, use spreadsheets, or prove theorems, the real utility of mathematics is not that you can use it to figure the slope of a wheelchair ramp, but that it provides you with the intellectual schemata necessary to make sense of a world in which the products of mathematical thinking are increasingly pervasive in almost every walk of life. This is not to say that other facets of mathematics should be neglected; questions of content, applications, cultural significance, and connections are all essential in the design of a mathematics program. But without explicit attention to mathematical ways of thinking, the goals of "intellectual sophistication" and "higher order thinking skills" will remain elusive.

When I first came to EDC in the early 1990s, my colleagues and I made a careful analysis of these mathematical habits of mind (see Cuoco, Goldenberg, & Mark [1996], for example), and we began developing high school courses and curricula organized around this analysis. *The CME Project* is a direct descendent of that early work and the decades of classroom experience that preceded it; the evolution is described in more detail in "The Case of *The CME Project*" (Cuoco, 2007).

By "mathematical habits of mind," I mean the mental habits that mathematicians use, often unconsciously, in their mathematical work. There are general mathematical habits—performing thought experiments, for example—and habits that are central to specific branches of mathematics. In analysis, for example, one often employs reasoning by continuity or passing to the limit. There are also important *algebraic* habits of mind that are the focus of the algebra courses in *The CME Project*. These include:

- Seeking regularity in repeated calculations.
- "Chunking" (changing variables in order to hide complexity).
- Reasoning about and picturing calculations and operations.
- Purposefully transforming and interpreting expressions to reveal hidden meaning.
- Seeking and modeling structural similarities in algebraic systems.

For example, in *The CME Project*, students see the identity

$$\left(\frac{a+b}{2}\right)^2 - \left(\frac{a-b}{2}\right)^2 = ab$$

in several different contexts. They learn to *imagine* expanding the left-hand side, without having to write anything down. And they *interpret* the identity in several different ways that connect algebra with geometry and optimization. For example, the identity can be used to establish the arithmetic-geometric mean inequality, and it can be used to show that a square maximizes the area for rectangles of fixed perimeter. At this conference, Tom Banchoff sketched a geometric proof of the identity.

Developing these and related algebraic habits is a pervasive goal throughout the program, especially in the two algebra courses. So, for example, *The CME Project* develops an approach to solving classical algebra word problems, not because of any intrinsic value in these problems and their stylized contexts, but because this class of problems, and the approach students use to solve them, provides an arena for developing the extremely useful habit of finding regularity in repeated calculations and forming processes from isolated computations.

Our choices of technologies and how we use them is also dictated by this goal of developing specific mathematical habits. For example, dynamic geometry environments can be used to help students learn to reason by continuity and to look for invariants under continuous transformations.

Computer algebra systems are ideal media for helping students develop algebraic habits like the ones described above. And access to a CAS gives students much more than computational power and the ability to perform complicated calculations.

USING A CAS TO BUILD ALGEBRAIC HABITS OF MIND

Modern CAS environments contain a great deal more than the ability to treat algebraic expressions as first-class objects. The TI-Nspire, for example, has graphics-handling capabilities (including equation graphing and dynamic geometry), a spreadsheet, a functional programming language, and a CAS, and all of these environments talk to each other. We make use of all of these capabilities in *The CME Project*, but I want to focus here on the value-added that comes from computer algebra: the ability to use these packages with formal algebraic expressions.

Our group at EDC sees three overlapping uses for computer algebra that help students develop algebraic habits: CAS media can be used as:

1. **an algebra laboratory.** CAS technology can be used to experiment with algebraic expressions in the same way that calculators can be used to experiment with numbers: generating data, making patterns apparent, and giving students the raw data from which they can generate conjectures. They provide teachers and students with general-purpose tools for finding regularity in data, or for imposing regularity when no simple patterns can be found. CAS technology also has the potential to bring a renewed and modern emphasis on formal algebra—that is, the algebra of forms—to school mathematics (see Godenberg [2003] and Cuoco and Levasseur [2003] for more on this theme).
2. **an algebraic calculator.** CAS technology can be used to make tractable and to enhance many beautiful classical topics, historically considered too technical for high school students. This is the use of technology that reduces computational overhead and that allows students to easily perform calculations that would be impossible (or overly distracting) without the technology. It is also the use that surrounds one of the biggest worries of many teachers in

the U.S.: If the computer can perform the calculations, what is the value of teaching paper-and-pencil algebraic skills?

3. **a modeling tool for algebraic structures.** CAS technology allows students to build models of algebraic objects that have no faithful physical counterparts. This use of technology adheres to our view that building a computational model for a mathematical structure helps one build the mental constructions needed to interiorize that structure (Cuoco and Goldenberg[1996], Harel and Papert [1991]). Furthermore, such computational models are *executable*, so that students can build working models of mathematical systems, turning the mathematician's thought experiments into actual experiments. What CAS environments add to other modeling environments is the facility to perform generic calculations with algebraic *expressions*—polynomials, rational functions, and formal power series. Since formal algebraic expressions are the "universal" objects in algebra, CAS environments provide a medium for expressing abstract algebraic structure.

Examples: A Case Study of $x^n - 1$

In this section, I'll look at each of the CAS uses described above—experimenting, calculating, and modeling—pointing out how they encourage the development of algebraic habits.

The context for these examples is the set of polynomials of the form x^n-1, where n is a positive integer. These polynomials are ubiquitous in almost every branch of mathematics. From a high school curriculum perspective, they can be used to tie together many core results from algebra, geometry, and trigonometry.

Experimenting: Finding factors of $x^n - 1$

Most first-year algebra books contain this factorization.

$$x^2 - 1 = (x-1)(x+1)$$

Sometime in high school, students may also see the following.

$$x^3 - 1 = (x-1)(x^2+x+1)$$

$$x^4 - 1 = (x-1)(x+1)(x^2+1)$$

$$x^6 - 1 = (x-1)(x+1)(x^2+x+1)(x^2-x+1)$$

So, over the integers $\mathbb{Z}$, $x^2 - 1$ and $x^3 - 1$ each have two factors, $x^4 - 1$ has three, and $x^6 - 1$ has four. Is there any pattern to the number of factors as a function of n? That is, can we find any regularity in this table?

n	Number of factors of $x^n - 1$
1	1
2	2
3	2
4	3
5	
6	4
7	
8	
9	

A CAS allows students to experiment with this question, generating data from which they can draw conclusions. For example, they can define a function that factors the polynomials (Screen 1):

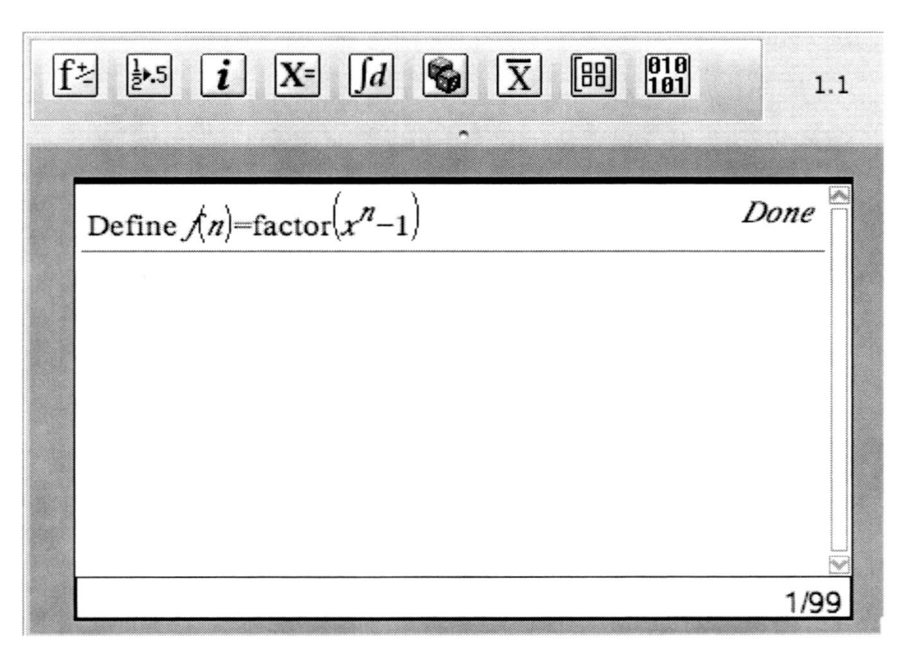

SCREEN 1.

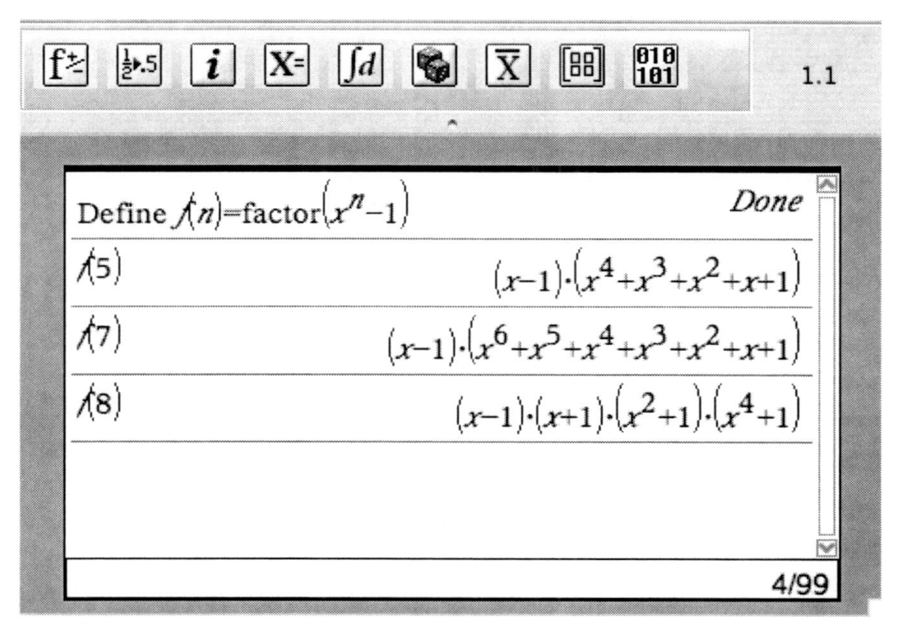

SCREEN 2.

The experiment might proceed as Screen 2. At this point, two conjectures often emerge:

1. There are always at least two factors:

$$x^n - 1 = (x - 1)(x^{n-1} + x^{n-2} + \cdots + x^2 + x + 1)$$

2. If n is odd, there are exactly two factors.

The first conjecture is true; the Factor Theorem found in almost any 2nd algebra course shows that $x-1$ must be a factor of $x^n - 1$ for any n, because 1 is a root of the equation $x^n - 1 = 0$. In *The CME Project*, we ask students to explain why the right-hand side multiplies out to $x^n - 1$ *without carrying out any explicit calculations*, picturing how the calculation would go if they did multiply everything out.

The identity

$$x^n - 1 = (x - 1)(x^{n-1} + x^{n-2} + \cdots + x^2 + x + 1)$$

is extremely useful in high school mathematics. It can be used, for example, to derive the formula for the sum of a geometric series, and it plays a role in helping students analyze the standard method for calculating the monthly payment on a loan.

Conjecture 2 is false, as a little more experimenting shows.

n	Number of factors of $x^n - 1$
1	1
2	1
3	2
4	2
5	2
6	4
7	2
8	4
9	3

When we've used this activity with students and teachers, several conjectures emerge:

- If n is *prime*, there are exactly two factors.
- If n is the square of a prime, there are three factors ($x^9 - 1$, for example).
- If n is the product of two distinct primes, there are four factors ($x^{15} - 1$, for example).

In classroom discussions or in student work, these statements usually coalesce into a single conjecture:

Conjecture: *The number of irreducible factors of $x^n - 1$ over $\mathbb{Z}$ is the number of positive integer factors of n.*

Here, students have come up with a conjecture for a non-obvious (and non-trivial) pattern in a sequence of polynomials. When I've used this activity with students and teachers, the question takes on a life of its own, and because of the laboratory environment afforded by the CAS, the objects of the investigation (the polynomials) become *real objects*. Some other points about this investigation:

- Even though the point of the activity is not to develop paper-and-pencil skill in factorization, it can be used to motivate the need for hand calculation by asking students how the CAS is coming up with its outputs. Many of the factorizations of $x^n - 1$ for small n *can* be

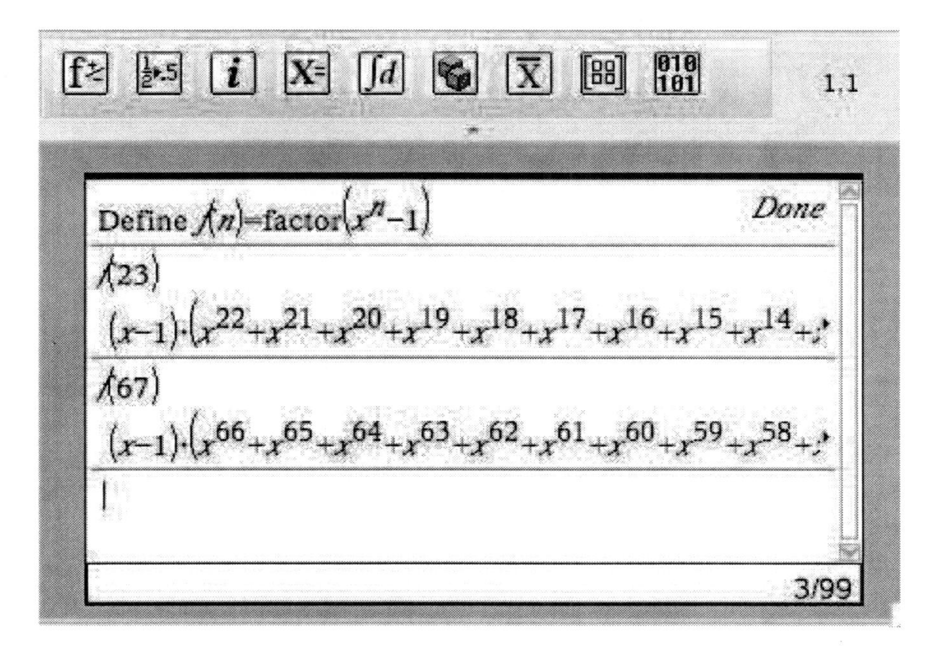

SCREEN 3.

carried out by hand, and a good follow-on to this investigation is to develop techniques for doing so.

- The CAS can be used to check conjectures for large values of n, adding to the sense that students are working with real objects (Screen 3).
- By looking at the actual factorizations produced by the CAS, rather than simply the number of factors, students can develop and prove more refined results. Indeed, the CAS can be used to inspire results about the factorizations of certain subsets of our sequence (Screen 4):
- This is a good example of a low-threshold–high-ceiling activity. CAS use makes progress on a conjecture tractable for almost all second-year algebra students, and many of them will leave it at that, or they may take things a bit further and show why $(x^n - 1)/(x - 1)$ is irreducible if n is prime. But there's much more to the story: the mathematics behind all this is central to many parts of algebra and analysis, and it gets deep enough to challenge even the most advanced students. Briefly, if (x) is the polynomial whose roots are precisely the primitive kth roots of unity, then:

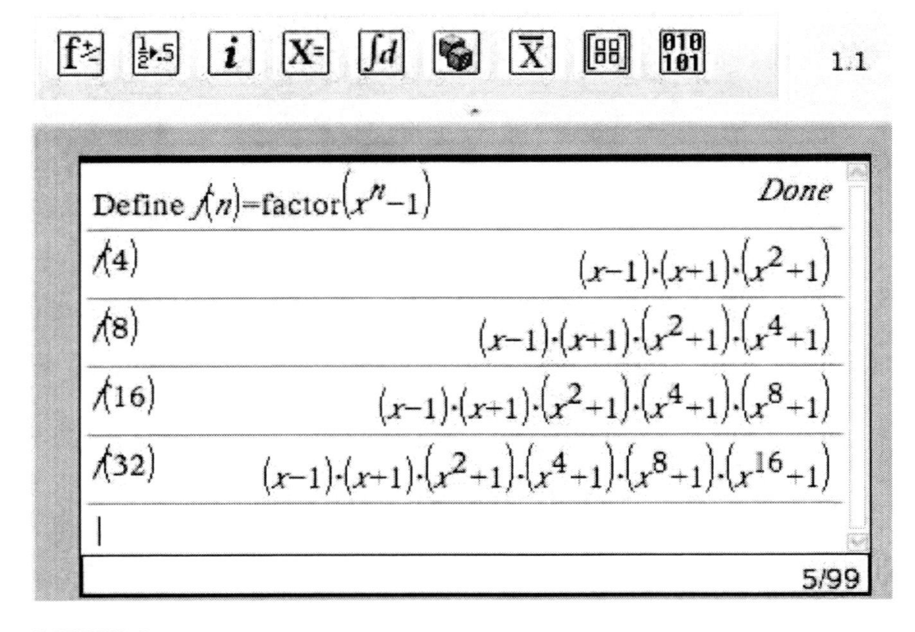

SCREEN 4.

$$x^n - 1 = \prod_{d|n} \psi_d(x) \qquad (*)$$

Here, the product is over all divisors of n. It can be shown (although the standard proof is quite hard in places) that each (x) is defined and irreducible over $\mathbb{Z}$ explaining why the conjecture of the students shown above is, in fact, true. Also, equation (*) can be used to compute each (x) recursively:

$$\psi_n(\chi) = \frac{\chi^n - 1}{\prod_{d\|n} \psi_d(\chi)}$$

where this time the product is over the proper divisors of n.

At this conference, someone asked if the coefficients of the (x) are always in the set

$$\{-1, 0, 1\}$$

One can use a CAS to investigate this question (Screens 5 and 6).

The first instance of a coefficient different from 0, 1, or –1 is in Ψ_{105} (105 is the first integer that is a product of three distinct odd primes). For more on the coefficients of the (x), see (Bloom [1968], Habermehl, Richardson, and Szwajkos, [1964]).

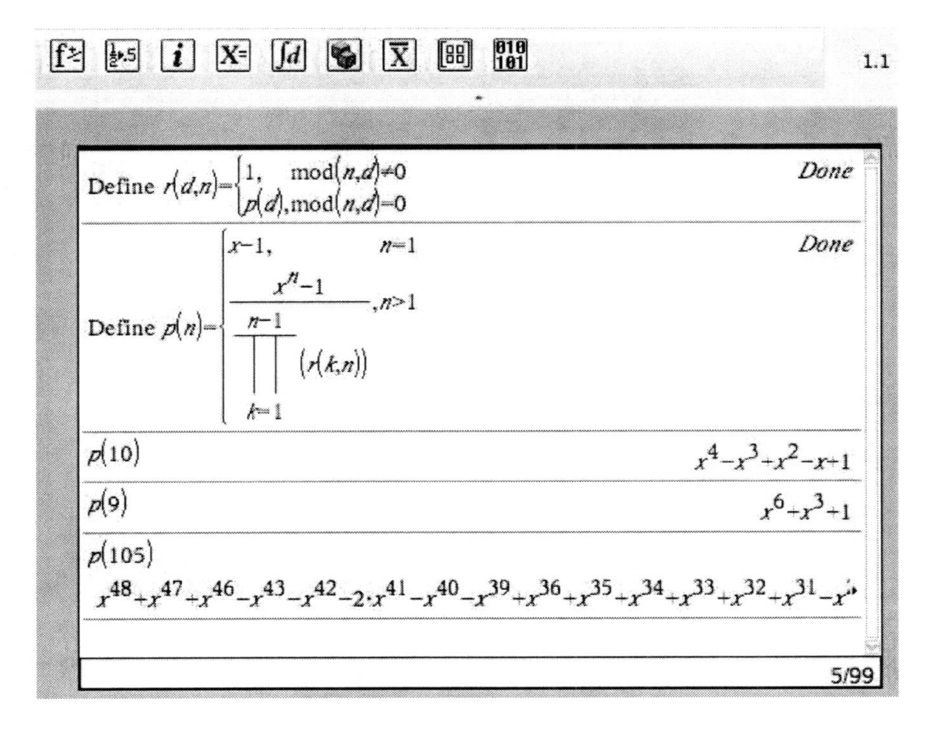

SCREEN 5.

REDUCING OVERHEAD: THE POLYNOMIAL FACTOR GAME

The *Connected Mathematics Project* (Lappan, Phillips, et.al.[2008]) introduces students to primes and the prime factorization of integers via the *Factor Game.* This is a game for two players, played on a board like this:

1	2	3	4	5
6	7	8	9	10
11	12	13	14	15
16	17	18	19	20
21	22	23	24	25
26	27	28	29	30

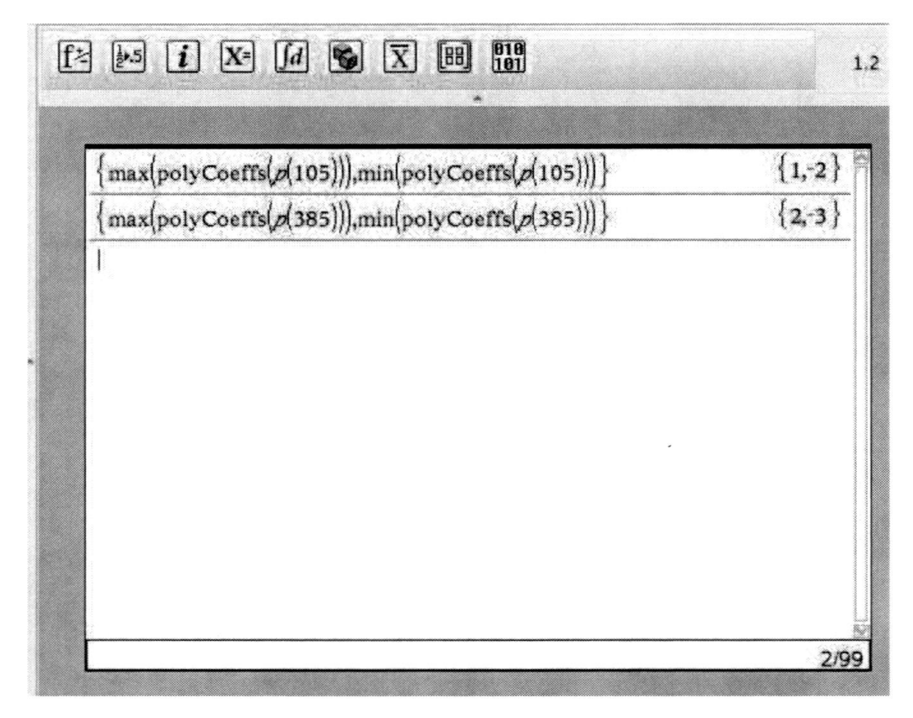

SCREEN 6.

The rules of the game are up for negotiation in a class, but one version goes like this:

1. Player A picks a number n from the board, getting that many points, and the number is crossed off.
2. Player B gets the sum of all the numbers not crossed off on the board that are factors of n, and crosses them off.
3. B goes next, picking an available number and gets that value.
4. A gets the sum of the non-crossed off numbers that are factors of m.
5. If either player picks a number with no factors left on the board, he or she loses a turn and gets no points.
6. The game continues until there are no possible moves.

The CME Project contains a game with the same rules, except the board looks like this:

$x-1$	x^2-1	x^3-1	x^4-1	x^5-1
x^6-1	x^7-1	x^8-1	x^9-1	$x^{10}-1$
$x^{11}-1$	$x^{12}-1$	$x^{13}-1$	$x^{14}-1$	$x^{15}-1$
$x^{16}-1$	$x^{17}-1$	$x^{18}-1$	$x^{19}-1$	$x^{20}-1$
$x^{21}-1$	$x^{22}-1$	$x^{23}-1$	$x^{24}-1$	$x^{25}-1$
$x^{26}-1$	$x^{27}-1$	$x^{28}-1$	$x^{29}-1$	$x^{30}-1$

The points that a player wins on a round correspond to the degrees of the polynomials that are picked.

The CAS is used here simply as an algebraic calculator. If a player wants to see if one of these polynomials divides another, he or she can simply check to see if the quotient is a polynomial (Screen 7).

It doesn't take long before students begin to see that this game "is the same as the middle school factor game." That is, a conjecture emerges

Conjecture*: x^m-1 is a factor of $x^n-1 \Leftrightarrow m$ is a factor of n*

One direction of this implication is a nice application of the "chunking" habit: To see, for example, that x^3-1 is a factor of x^4-1, you can argue like this:

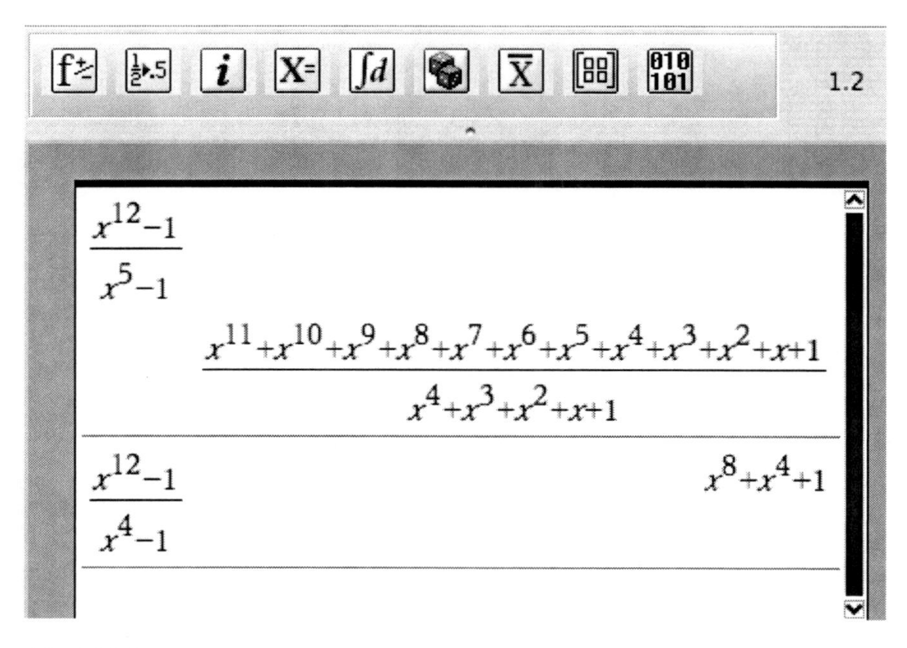

SCREEN 7.

$$\begin{aligned} x^{12} - 1 &= (x^3)^4 - 1 \\ &= (♣)^4 - 1 \\ &= (♣ - 1)\,(♣^3 + ♣^2 + ♣ + 1) \qquad (**) \\ &= (x^3 - 1)((x^3)^3 + (x^3)^2 + (x^3) + 1) \\ &= (x^3 - 1)\,(x^9 + x^6 + x^3 + 1) \end{aligned}$$

The other direction of the implication (if $x^m - 1$ is a factor of $x^n - 1$, then m is a factor of n) is much harder. One way to think about it requires some facility with De Moivre's theorem and with roots of unity. At this conference, Vince Matsko showed me another way to derive this implication that doesn't use complex numbers. Briefly, it goes like this:

Suppose that $x^m - 1$ is a factor of $x^n - 1$. Write $n = mq + r$ with $0 \le r < m$. Then:

$$x^{n-r} - 1 = x^{qm} - 1$$

But $x^m - 1$ is a factor of the right-hand side of this equation (chunking, again), so it divides both $x^n - 1$ and $x^{n-r} - 1$, and hence divides their difference:

$$x^{n-r}(x^r - 1)$$

But $x^m - 1$ is relatively prime to x^{n-r}, so it must be a factor of $x^r - 1$. Since $r < m$, this implies that $r = 0$.

MODELING: ROOTS OF UNITY

If you watch high school students calculate with complex numbers, many will act as if they are calculating with polynomials in i, with the additional simplification rule "$i^2 = -1$." There is a germ of an important idea here: students are noticing the structural similarities between $\mathbb{C}$ and $\mathbb{R}[x]$—the two systems seem to "calculate the same." This seeking structural similarities in algebraic systems is an important algebraic habit of mind, and it gets exercised when calculations in one system start to feel like calculations in another.

Many precalculus courses (including *The CME Project*) contain a treatment of De Moivre's theorem, often stated like this:

$$(\cos\theta + i\sin\theta)^n = \cos n\theta + i\sin n\theta$$

De Moivre's Theorem implies several facts relevant to our family $x^n - 1$:

- The roots of $x^n - 1 = 0$ are

$$\left\{\cos\left(\frac{2k\pi}{n}\right)+i\sin\left(\frac{2k\pi}{n}\right):\ 0 \le k < n\right\}$$

- If

$$\zeta = \cos\left(\frac{2\pi}{n}\right)+i\sin\left(\frac{2\pi}{n}\right)$$

  these roots are $1, \zeta, \zeta^2, \zeta^3,, \ldots, \zeta^{n-1}$.
- These roots lie on the vertices of a regular n-gon of inscribed in the unit circle in the complex plane.

In *The CME Project* precalculus book, an optional suite of problems deals with the 7th roots of unity (Figure 1).

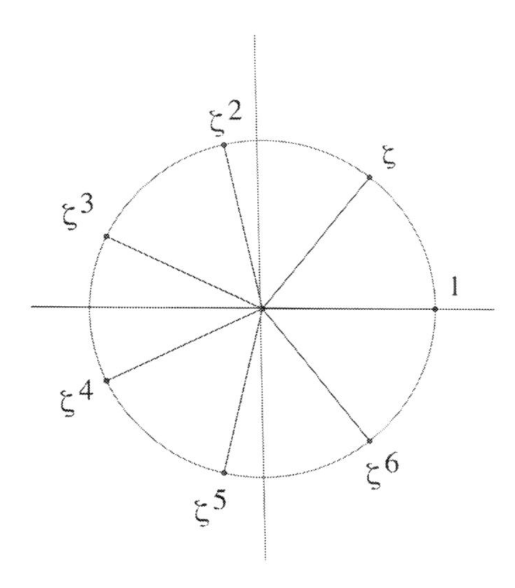

FIGURE 1.

Notice that

- The six non-real roots come in conjugate pairs.
- So, $(\zeta + \zeta^6)$, $(\zeta^2 + \zeta^5)$, and $(\zeta^3 + \zeta^4)$ are real numbers.
- Hence these three numbers satisfy a cubic equation over R.

What is it? Let $\alpha = \zeta + \zeta^6$

$$\beta = \zeta^2 + \zeta^5$$

$$\gamma = \zeta^3 + \zeta^4$$

To find an equation satisfied by α, β, and γ, we need to find

- $\alpha + \beta + \gamma$
- $\alpha\beta + \alpha\gamma + \beta\gamma$
- $\alpha\beta\gamma$

We find these one at a time. . .

The Sum: From their definitions,

$$\alpha + \beta + \gamma = \zeta^6 + \zeta^5 + \zeta^4 + \zeta^3 + \zeta^2 + \zeta$$

But

$$x^7 - 1 = (x - 1)(x^6 + x^5 + x^4 + x^3 + x^2 + x + 1)$$

So,

$$\zeta^6 + \zeta^5 + \zeta^4 + \zeta^3 + \zeta^2 + \zeta = -1$$

The Product:

$$\alpha\beta\gamma = (\zeta + \zeta^6)(\zeta^2 + \zeta^5)(\zeta^3 + \zeta^4)$$

We can get the *form* of the expansion by expanding

$$(x + x^6)(x^2 + x^5)(x^3 + x^4)$$

A CAS tells us that

$$(x + x^6)(x^2 + x^5)(x^3 + x^4) =$$

$$x^{15} + x^{14} + x^{12} + x^{11} + x^{10} + x^9 + x^7 + x^6$$

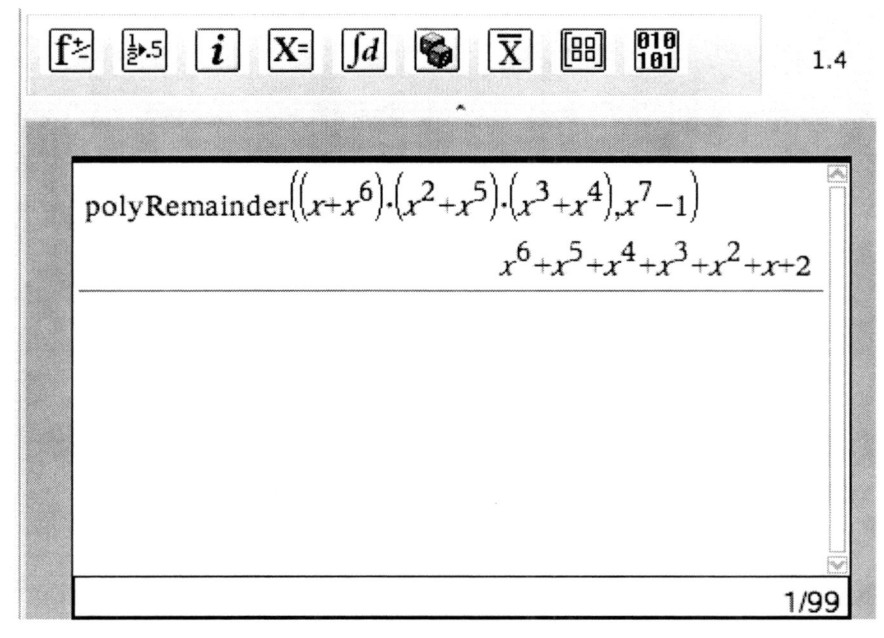

SCREEN 8.

But if we replace x by ζ , we can replace x^7 by 1. So, if the above expression is written as:

$$(x^7 - 1)\,q(x) + r(x)$$

Replacing x by ζ will produce $r(\zeta)$. A CAS can be used to do the calculation (Screen 8). Since

$$\zeta^6 + \zeta^5 + \zeta^4 + \zeta^3 + \zeta^2 + \zeta + 1 = 0$$

We get

$$\alpha\beta\gamma = 1$$

The sum, two at a time: Well,

$$\alpha\beta + \alpha\gamma + \beta\gamma = (\zeta + \zeta^6)(\zeta^2 + \zeta^5) + (\zeta + \zeta^6)\ (\zeta^3 + \zeta^4) + (\zeta^2 + \zeta^5)(\zeta^3 + \zeta^4)$$

We can use a CAS, thinking of this as a formal calculation, reducing by $x^7 - 1$ (Screen 9):

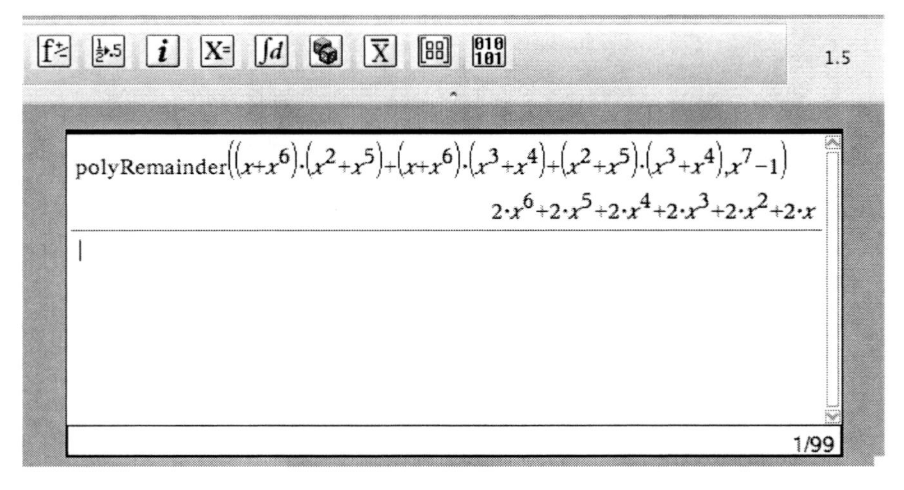

SCREEN 9.

It follows that $\alpha\beta + \alpha\gamma + \beta\gamma = -2$, and our cubic is

$$x^3 + x^2 - 2x - 1 = 0$$

There are several purposes for this exercise:

- In an informal way, students preview the idea that one can model $\mathbb{Q}(\zeta)$ by "remainder arithmetic" in $\mathbb{Q}(x)$, using $x^7 - 1$ as a divisor.

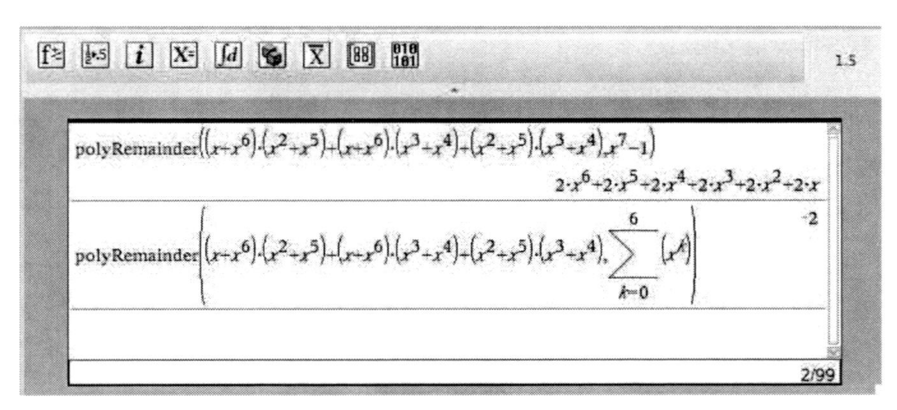

SCREEN 10.

- In fact, one can use any polynomial that has ζ as a zero—the one of smallest degree is

$$x^6 + x^5 + x^4 + x^3 + x^2 + x + 1$$

Doing so would have reduced significantly the simplifications needed at the end of each step, and the CAS would carry out the calculations just as easily.
- This previews Kronecker's construction of splitting fields for algebraic equations.

CONCLUSION

CAS environments have been used for over a decade in undergraduate mathematics, and now, with the availability of these media on handheld devices, they are gradually making their way into precollege (upper secondary) programs. Especially in the United States, where jumping on bandwagons has a longstanding and quasi-respectable tradition in education, two opposing camps are developing:

- In one camp are those worried that the influx of CAS environments into precollege mathematics will produce a generation of high school students who reach for a calculator to factor $x^2 + x$, much like the alleged current generation of college students who reach for a calculator to multiply 57 by 10.
- In the other camp are those who adopt the motto "if the machine can do it, why bother teaching it?"—many educators are proclaiming that facility with algebraic calculation is unnecessary and that we can do away with those tortuous pages of factoring, simplifying, and solving.

Experience tells us that both of these extreme stances will evolve eventually into something much less grandiose and that CAS environments will take their place alongside other useful computational media as enhancements to, rather than replacements for, the essential role that technical fluency plays in mathematical understanding.

In this paper, I've provided one example of how CAS environments can be used to enhance the high school algebra curriculum. *The CME Project* uses CAS technology to

1. *experiment with algebra*
2. *reduce computational overhead*
3. *use polynomials as modeling tools.*

The examples given in the previous section are just that: examples. There are many other examples of enhancements that have little to do with x^n-1: Chebyshev polynomials, Lagrange interpolation, Newton's difference formula, and generating functions, just to name a few. In the age of CAS, all of this beautiful and classical mathematics becomes accessible to many more students than in previous decades.

REFERENCES

Bloom, D. M. (1968). On the coefficients of the cyclotomic polynomials. *The American Mathematical Monthly, 75*(4), 372–377.

Cuoco, A. (1998). Mathematics as a way of thinking about things. In L. Steen, (Ed.), *High school mathematics at work.* Washington, D.C.: National Academy Press.

Cuoco, A., & Goldenberg, P. (1996). A role for technology in mathematics education. *Journal of Education, 178*(2), 15–32.

Cuoco, A., Goldenberg, P., & Mark, J. (1996). Habits of mind: An organizing principle for mathematics curriculum. *Journal of Mathematical Behavior, 15*(4), 375–402.

Cuoco, A., & Levasseur, K. (2003). Classical mathematics in the age of CAS. In J. Fey, A. Cuoco, C. Kieran, L. McMullin, & R.M. Zbiek (Eds.), *Computer Algebra Systems.* (pp. 97–116). Reston, VA: NCTM.

Cuoco, A. (2007). The Case of *The CME Project.* In C. Hirsch (Ed.), *Perspectives on Design and Development of School Mathematics Curricula.* Reston, VA: NCTM.

Education Development Center (2009). *The CME Project.* Boston, MA: Pearson.

Goldenberg, E. P. (2003). Algebra and computer algebra. In J. Fey, A. Cuoco, C. Kieran, L. McMullin, & R.M. Zbiek (Eds.), *Computer Algebra Systems* (pp. 9–31). Reston, VA: NCTM.

Habermehl, H., Richardson, S., & Szwajkos, M.A. (1964). A note on coefficients of cyclotomic polynomials", *Mathematics Magazine, 37*(3), 183–185.

Harel, I. & Papert, S. (1991). *Constructionism.* Norwood, NJ: Ablex Publishing.

Lappan, Glenda, Elizabeth Phillips, et al. (2008). "Prime Time", *CMPII.* Boston, MA: Pearson.

CHAPTER 9

A PERSPECTIVE ON THE FUTURE OF COMPUTER ALGEBRA SYSTEMS IN SCHOOL ALGEBRA

M. Kathleen Heid

This paper gives one person's perspective on computer algebra in mathematics education. The examples are drawn from my experience, and the story is told from my perspective. The references are mainly ones that explain my perspective and tell my story. Others would draw on different resources, tell different stories, and give quite different insights.

BACKGROUND: CAS 25 YEARS AGO

Over the past twenty-five years, I have been involved in research and development centered on the use of computer algebra systems (CAS). My first attempt occurred in 1982-1983 soon after the release of muMath, the grandfather of Derive. The first symbolic manipulation program incorporated on Texas Instruments calculators was a derivative of Derive. MuMath was the first symbolic manipulation system available for what we called a microcomputer at that time. At that time, there were no integrated symbolic-graphical-numeric programs – programs the field would grow to call

Future Curricular Trends in School Algebra and Geometry: Proceedings of a Conference. pages 129–143

computer algebra systems. In the context of his NSF project to examine the role of technology in school mathematics, my doctoral advisor (Jim Fey) had purchased copies and we took on the challenge of thinking about how muMath (and programs like it) could be used in mathematics instruction. The authoring team for muMath, led by Albert Rich and David Stoutemyer, took as its mission adapting the MACSYMA program so that it could work using the minimal memory then available on microcomputers. The authors called their program muMath because the letter mu was used to represent micro in the metric system and this program was designed to run on microcomputers. muMath-80 ran on the Apple II with a special memory-boosting card (that increased its capacity to 64K). It was fascinating to me at the time since, as I indicated in a *Computing Teacher* article (Heid, 1983), it would receive about a B on the standard calculus exam of that time. It generated the exact form of derivatives and integrals, but it did not generate the constants in integration. It was also intriguing because it allowed the user to control the form of the output – not only could it conduct symbolic transformations to produce equivalent forms but the user could set control variables to determine which equivalent form to use. Unfortunately, current computer algebra systems do not allow the user this option.

BEGINNING TO EXPLORE CAS IN MATHEMATICS EDUCATION: CALCULUS

I identified calculus as an area that seemed unduly focused on symbolic manipulation alone, and I wondered how incorporating technology including technology with graphical, tabular/numeric, and symbolic manipulation capability would affect the understandings that students typically developed in courses like these. It was clear that the Mathematics Department would not tolerate a revised version of their engineering calculus course, so I took on the challenge of reformulating a business calculus course. The task I took on was to use muMath and other programs to reformulate a calculus course so that it could focus on concepts instead of the typical focus on helping students learn to perform symbolic manipulations accurately and quickly (Heid, 1984, 1988). I supplemented muMath with homegrown graphing and table programs—so my students were equipped with the 1982 equivalent of a computer algebra system. I created a course that, for the first 12 weeks, focused on developing students' conceptual understanding while using the computing facilities to generate graphs, tables, and symbolic manipulation. I spent the last three weeks of the class introducing students to the symbolic procedures on which the comparison class spent the semester. The result was that, when compared with students in a typical calculus course (taught by an experienced and successful professor), students developed superior conceptual understanding and performed comparably to the comparison class on measures of by-hand symbolic manipulation. One conclusion that might be drawn is that either calculus is really a three-week course, or there is something about developing conceptual understanding that enables quicker acquisition of symbolic skills. I thought about this as the conceptual understanding forming a framework for students on which to hang the symbolic manipulation skills they later developed.

A FOLLOW-UP: CAS IN BEGINNING ALGEBRA

We had shown that configurations of software somewhat equivalent to computer algebra systems could be used to reformulate introductory calculus to focus on concept development instead of solely on performance of procedures. But the students in introductory calculus entered the course having completed at least two years of high school algebra. It was not clear whether there was something about this background that enabled students' development of conceptual understanding. If computer algebra systems were to have a real effect on school mathematics, they needed to be tested with students who had not had previous exposure to algebraic symbolic manipulation. The question was whether an approach like this would work with beginning secondary school algebra students. Jim Fey made a successful proposal to the National Science Foundation for the creation of technology-based modules for such a beginning algebra course. He invited me to be part of that team, and the team produced several modules that introduced students to algebra using graphical, numerical, and symbolic software. This work continued in a subsequent project, for which I served as co-PI, to create a complete technology-intensive beginning algebra course—the result was *Computer-Intensive Algebra* (Fey et al., 1991), later called *Concepts in Algebra: A Technological Approach* (Fey et al., 1995, 1999) and known as CIA. CIA was conceived over a six-year period and born in 1991. The course was functions-based, applications-oriented, and conceptually driven. Students focused on using computer-generated graphs, tables, and symbols to analyze linear and quadratic real-world based situations. Students had access to the computer algebra software throughout the course, and the final two chapters focused their attention on symbolic reasoning—engaging them in thinking about the meaning of equivalent equations and equivalent expressions (graphically, numerically, and symbolically). We took seriously the notion of investigating the effects of the curriculum, and conducted research on the conceptual and procedural understandings that students developed.

RESEARCH ON CAS IN BEGINNING ALGEBRA

One study we conducted (Heid, 1996) involved students who were tracked between those who were taking their first algebra course in eighth grade and those who were taking their first algebra course in ninth grade. Students in the CIA classes began their study of algebra midway through eighth grade and concluded their work with the CIA curriculum April 6 of their ninth-grade year (two months before the end of the school year). They studied by-hand symbolic manipulation for the remainder of the year and completed a final exam that focused on by-hand symbolic manipulation skills. We investigated their conceptual and procedural skills and understandings and compared it to the skills and understandings of the eighth-grade and ninth-grade algebra students. The results of the study paralleled those of the calculus study. Students in the CIA classes scored better than the ninth-grade students on every final exam item and as good as or slightly less than the eighth-grade students on the same final exam items. In other words, the level of their procedural skills was as expected (between the students in the advanced track and those in the regular track) after only six weeks of work on these skills. On every measure of conceptual understanding, the CIA students showed better conceptual understanding than their peers in the other algebra classes. We had garnered evidence that a technology-intensive introductory algebra course could generate better conceptual understanding and equivalent procedural skills. A remaining question was what accounted for those differences. Students had access to a new curriculum and to technology with graphing, numerical, and symbolic manipulation capabilities. It was not possible to tease out a particular influence of the symbolic manipulation software.

EXTENDING CAS TO SECONDARY MATHEMATICS BEYOND BEGINNING ALGEBRA

Our work continued in developing technology-intensive curricula. In a project that Rose Zbiek and I co-directed, we developed nine modules of technology-intensive materials that could be used with students who had the equivalent of a CIA algebra background (Heid & Zbiek, 2003). Students were to have access to computer algebra systems and other software throughout their use of the materials. We created modules on a range of mathematical topics, some of which were natural territory for computer algebra systems. Our concluding module was focused exclusively on using CAS to enhance symbolic reasoning.

ISSUES OF IMPLEMENTATION AND SCALING UP

Throughout the years of piloting CIA and then CAS-IM in classrooms throughout the United States, we struggled with the problems of scaling up. We wondered what it would take for computer algebra systems to be embraced in a broad range of schools. We addressed this concern following the publication of the CIA curriculum. Through an NSF project, we (Glen Blume was co-PI) involved teachers in 30 states from a broad range of settings (urban, suburban, rural, high-minority, low-minority, magnet schools, regular schools) in implementing the CIA curriculum. We conducted a four-week residential on-campus workshop for teachers, introducing them to using computer algebra systems to teach introductory algebra (Heid, Zbiek, & Blume, 1997). At the time (the mid 1990's) it was the rare Algebra I teacher who had had any experience with a computer algebra system. These teachers returned to their schools to implement the CIA curriculum. For the most part they were quite successful in using the curriculum with students at a range of levels, but many struggled with their school districts in convincing them that this technology-based, CAS-active course was preparing students with adequate procedural skills. One teacher, for example, who taught in an inner-city high school, told stories of her struggle and success with CIA students. She told of her surprise one day when her students were arguing vehemently in one corner of the room; she was preparing herself to break up a serious fight when nearing the site of the argument she realized that the students were arguing about how best to conduct a mathematics experiment. She described this steady focus on mathematics as unusual for students in other classes but usual for her CIA students. She also told about students who had finished the CIA curriculum and gone on to Algebra II. Their Algebra II teacher began the year complaining to this teacher that the CIA students could not do certain manipulative skills. As the year went on, however, that same teacher returned to the CIA teacher, remarking that the CIA students outshined their non-CIA peers in understanding and solving word problems. In spite of indications of success such as these, it was difficult to locate schools that would try the CIA and CAS-IM curricula. That difficulty is compounded today in the era of No Child Left Behind, with its laser focus on testing procedures.

CURRENT EMPHASIS AND TREATMENT OF ALGEBRA USING CAS

Many teachers were and are hesitant to bring into their classrooms calculators that perform the same manipulative procedures that they have spent their careers developing in students (Freda, 2008; Heid, 2002). Without curricula designed to capitalize on computer algebra systems, it is hard for many teachers to envision what they would teach if students can use calculators to perform symbolic manipulation. Especially in the era of high-stakes testing precipitated by legislation such as No Child Left Behind, school systems provide an atmosphere in which teachers feel the need to teach directly to the procedures and problem types identified by their school systems. In many school systems, there seems to be no room for deviating from a curriculum focused on testing.

Computer algebra systems afford teachers opportunities to focus student attention on the concepts of algebra (Heid, 1996), on big mathematical ideas instead of on the details of procedures (Zbiek & Heid, in press), and on more general and abstract strategies (Heid, 2003). If schools are to capitalize on the availability of computer algebra systems, there is a need for curricula that embrace computer algebra systems, for development in teachers of confidence that skills can develop without focusing an entire curriculum on them, and for teachers and schools that are willing to take the risk of rethinking the goals of school mathematics in light of available technology. And there is a need for the Federal government to recognize the importance of schools embracing technology and leading the charge to 21st Century school mathematics. This constellation of needs is far from being met.

One of the reasons given for not using technology is the belief that there is no research to support their use. In response to these concerns, Glen Blume and I edited two volumes whose chapters were written by internationally-known researchers whose work had influenced the use of technology in mathematics education. These volumes synthesized research on technology and the teaching and learning of mathematics, provided cases of the use of the technology in the teaching of mathematics, and offered perspectives on cross-cutting issues (Blume & Heid, 2008; Heid & Blume, 2008a). Some of the research on use of CAS in school mathematics is addressed in the synthesis of research on algebra (Heid & Blume, 2008b). Research on CAS in U.S. classrooms, however, is sparse. Our colleagues in Australia, Austria, France, and Great Britain, however, have conducted seminal research on CAS in classrooms, developing theory about their use as well as classroom-based examples. One particularly insightful book developed by international editors focuses on a theoretical perspective on the use of symbolic manipulation calculators in school mathematics (Guin, Ruthven, & Trouche, 2005). The *International Journal for Technology in Mathematics Education,* formerly known as the *International Journal for Computer Algebra in Mathematics Education* has in the past focused entirely on computer algebra in mathematics instruction, and the *International Journal of Computers for Mathematical Learning* has published seminal research on CAS in mathematics classrooms. Although U.S. research on the use of CAS is minimal, international researchers have been focusing on the impact of CAS in mathematics teaching. Because of that research, we know more about teachers' use of CAS (Monaghan, 1997), about how teachers' preferences in use of CAS influence what students learn (Kendal & Stacey, 2001), and how effective use of CAS requires that teachers and students develop a relationship with the tool that changes both the tool and what the user does with it (Lagrange, 1999; Trouche, 2004). Much of the existing research has focused on students in their final years of secondary school, and more research is needed, especially on the use of CAS with younger secondary students.

In order to enable research that translates well into practice, teachers, curriculum developers, and researchers need to collaborate in establishing venues for that research. More groups of teachers such as the Chicago-area MEECAS group[1] need to take on the challenge of collaborating with researchers in investigating the effects of the use of CAS in school mathematics. The purpose of the MEECAS group appears on their website:

> The purposes of MEECAS are to encourage active interest in the use of computer algebra systems and related technology in schools, to improve learning of mathematics, to influence curriculum, assessment, and pedagogy, to promote teacher development and experimentation with computer algebra systems, to support research in computer algebra systems in education, and to provide opportunities for exchange of ideas about teaching and learning with computer algebra systems. (Mathematics Educators Exploring Computer Algebra Systems (MEECAS))[1]

The year-long activities of the MEECAS group place cutting-edge teachers in contact with each other exploring and experimenting with the use of CAS in their classrooms, and the annual USACAS meeting has introduced CAS-using teachers in the United States to CAS researchers worldwide.

THE ROLE OF RESEARCH AND AVAILABILITY OF TECHNOLOGY IN MY WORK

As suggested in the introduction, research has been the guiding force in my work with technology. The curricula on which I have worked were all premised on some feature we investigated. In the calculus curriculum and in CIA, we investigated the impact on students' understanding of the mathematics of using CAS to focus on the concept of function. We examined whether it was feasible to shift procedural work to CAS in order to enhance conceptual understanding. In the CAS-IM curriculum, we investigated the ways in which students use representation in a multirepresentational environment[2] (Blume et al., 2002; Foletta, 2002; Heid et al., 2002; Zbiek, 2002a; Zbiek, 2002b; Zbiek & Finken, 2002). In the CIME work, we investigated in a CAS environment how teachers' concept of what learning is influenced how they assessed student learning (Heid, Blume, Zbiek, & Edwards, 1998/1999). In another experiment we investigated the feasibility of beginning algebra instruction with word problems when students had access to a symbolic manipulation program (Heid & Kunkle, 1988). In another, we investigate the feasibility of using CAS with students who were tracked at the very lowest level.

Availability of technology has impacted my work in several ways. First, there is a synergistic relationship between the technology that exists and

[1]Quote taken from www.meecas.org accessed May 1, 2008.

[2]Papers on this issue are posted on www.nku.edu/~foletta/ accessed May 1, 2008.

curriculum experiments with it. Because different tools afford the opportunity to pursue different mathematics, there are different issues to research. Second, the availability of tools in the schools makes a tremendous difference. This difference is related to how familiar teachers are with the technology—the stage of their instrumental genesis (Guin, Ruthven, & Trouche, 2005) with the CAS. Third, on a related issue, what counts is not only the technology that is available but the curriculum that is constructed to take advantage of that technology.

IMPLICATIONS FOR THE DEVELOPMENT OF MATHEMATICAL REASONING AND PROOF IN STUDENTS

I would like to paint a picture of the use of CAS in the development of mathematical reasoning and proof in students by recapping a recording of one talented seventh grader's work in beginning algebra using a computer algebra system (also see Iseri, 2003 and Heid, Hollebrands, & Iseri, 2002). At the time of the interview, Kevin's only school-based introduction to algebra was through the CIA materials (*Concepts in Algebra: A Technological Approach*). The problem that Kevin is considering is the following (these data were gathered in the context of Linda Iseri's (2003) dissertation).

Kevin readily generated a function rule:

$$y = \frac{x - \frac{1}{2}}{x + 3}$$

Create a function, k, that meets the following constraints:

The function is undefined at -3.

The function has a zero at $\frac{1}{2}$.

The function is always non-negative.

The function's domain is $[-5, \infty) - \{-3\}$

The function contains the point (4, 7)

Start by writing a function that meets the first constraint. Then I'd like you to alter the function rule so that it meets the second constraint, while still meeting the first constraint, and so on.

FIGURE 1.

that is consistent with the first two constraints. When faced with the next constraint, however, Kevin made several strategic guesses. He first thought he could produce non-negative values by squaring the numerator of his existing rule, thinking that the act of squaring would make the output positive (the rule he generated was:

$$y = \frac{x^2 - x + \frac{1}{4}}{x+3}.$$

He quickly realized that even when the numerator was positive, the denominator could still be negative. He then conjectured that generating a rule that was a cubic divided by a linear function would result in a square function–a function whose values he thought would all be positive. The rule he generated (by cubing the numerator) was:

$$y = \frac{x^3 - \frac{3}{2}x + \frac{3}{4}x - \frac{1}{8}}{x+3}.$$

Moving back and forth from rule to graph to adjusted rule to adjusted graph, Kevin saw that the graph for his new rule appeared as shown in Figure 2 and had some negative values.

Upon a prompt by the teacher to think of the rule as a single entity, Kevin generated the rule

$$y = \left|\frac{x - 0.5}{x+3}\right|$$

succeeding in fulfilling the first three constraints. After a few other conjectures, Kevin adjusted his rule to accommodate the fourth constraint, producing the rule

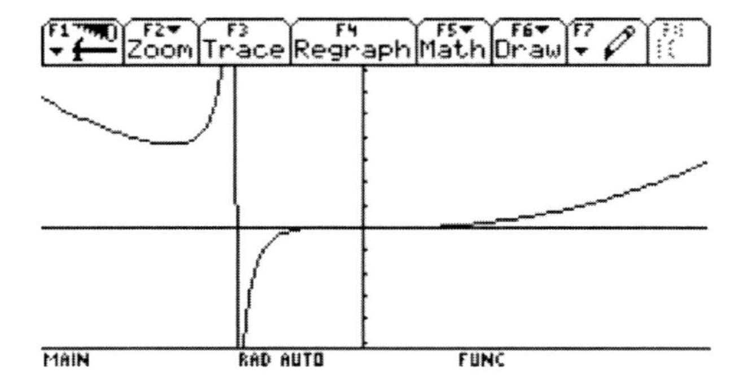

FIGURE 2.

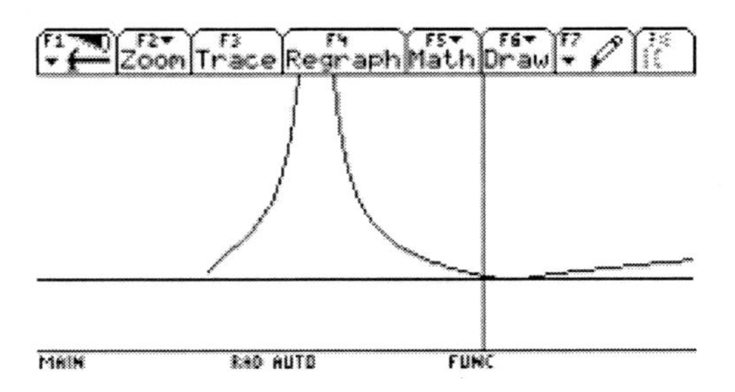

FIGURE 3.

$$y = \sqrt{x+5}\left|\frac{x-0.5}{x+3}\right|,$$

a rule with the graph shown in Figure 3.

The final constraint, that the function contain the point (4,7), posed a series of additional challenges for Kevin. At first he decided that he could accomplish his objective by adding a linear function rule that is 0 at $x = ½$ and that adds enough at $x = 4$ to make the corresponding y-value 7. He generated this rule,

$$y = \frac{11}{7}x + \frac{11}{14},$$

added it to his current rule to obtain

$$y = \sqrt{x+5}\left|\frac{x-0.5}{x+3}\right| + \frac{11}{7}x - \frac{11}{14},$$

and saw that he had generated a graph (Figure 4) that had some negative output values, violating the third constraint.

Having used the absolute value function before to generate non-negative values, Kevin "absolute-valutized" (his word) the rule to produce

$$y = \left|\sqrt{x+5}\left|\frac{x-0.5}{x+3}\right| + \frac{11}{7}x - \frac{11}{14}\right|.$$

This time, however, the apparent domain of the function changed to the real numbers except -3 since the calculator was (unbeknownst to Kevin) set on complex number mode. To circumvent this new twist, Kevin changed

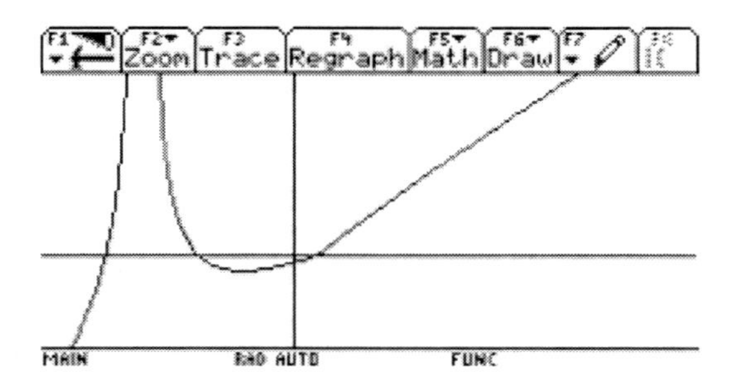

FIGURE 4.

the position of the absolute value sign, generating a rule whose graph has no negative outputs but that no longer contains (4, 7). He solved this problem by dividing by 18/7, generating a final function rule,

$$y = \frac{\left|\sqrt{x+5}\left|\frac{x-0.5}{x+3}\right| + \frac{11}{7}x - \frac{11}{14}\right|}{\frac{18}{7}},$$

that satisfied all constraints.

Kevin used the CAS when he needed to – and he floated fluidly between and among representations as he solved problems. Because the CAS displayed and operated on exact function rules, Kevin was able to identify and manipulate chunks of varying sizes to accomplish his objectives. While Kevin's facility with thinking of different portions of the rule as objects that can be manipulated is not likely for a large number of students, the point is that the CAS provides a venue in which students are afforded the opportunity to manipulate function rules and parts of function rules. With CAS tools, an appropriate curriculum, and a teacher prepared to take advantage of the tools and curriculum, student activity can focus on more structural features of symbolic manipulation.

IMPLICATIONS FOR THE PRE-SERVICE AND INSERVICE EDUCATION OF MATHEMATICS TEACHERS

The incorporation of CAS into the school mathematics curriculum in ways that capitalize on the symbolic manipulation capacity of CAS requires an in-service or professional development program tailored to enhancing teachers' understanding of the use of technology in the teaching and learning of mathematics. At Penn State, prospective teachers complete one or two courses that include specific foci on the use of CAS technology in the development of mathematical ideas. A course on teaching with technology uses CAS as one of several major tools for examining how to incorporate technology in secondary mathematics teaching. A course on functions capitalizes on CAS to enhance prospective teachers' understandings of function, variable, parameter, and equivalence. We also offer a graduate-level course on research on technology in the teaching and learning of mathematics that engages graduate students in examining and conducting research on major constructs that underpin the teaching and learning of mathematics.

CAS IN ALGEBRA AND THE NEXT 10 YEARS

The past few years and the coming years have witnessed several important curriculum development projects that have made an expressed effort to use CAS to various extents. Examples include the UCSMP third edition (Brown et al., 2008), the CME Project from the Education Development Center (CME, 2009a, 2009b), and the CD from our previously referenced Technology-Intensive Mathematics (Heid et al., 2004). It is my hope that teachers and schools, in trying these curricula, become convinced of their viability and spread the word to other teachers and schools. I would like to see more research on how teachers and students use CAS in curricular settings in which CAS use is assumed. I would like to see more research on what teachers and students learn in using CAS in curricular settings in which CAS use is assumed. And given that all of that happens, I would like to see curriculum developers take the lead in creating curricula that are based on the assumption that students will be using the tools of the 21st Century and that work to develop in students the kind of sophistication we see in students like Kevin.

REFERENCES

Blume, G. W., & Heid, M. K. (Eds.). (2008). *Research on technology in the learning and teaching of mathematics Vol. 2, Cases and perspectives.* Charlotte, NC: Information Age.

Blume, G., Heid, M. K., Hollebrands, K., & Piez, C. (2002, April). *Patterns in students' engagement with tasks in a technology-intensive secondary school mathematics curriculum.* Paper presented at the Research Presession to the 2002 Annual Meeting of the National Council of Teachers of Mathematics. Paper retrieved May 1, 2008, from www.nku.edu/~foletta/

Brown, S. A., Breunlin, R. J., Wiltjer, M. H., Degner, K. M., Eddins, S., K., Edwards, et al. (2008). *Algebra.* (3rd ed.). Chicago: Wright Group/McGraw-Hill.

CME Project. (2009a). *Algebra 1.* Pearson Education, Inc.

CME Project. (2009b). *Algebra 2.* Pearson Education, Inc.

Fey, J. T., & Heid, M. K., with Good, R. A., Sheets, C., Blume, G. W., & Zbiek, R. M. (1995, 1999). *Concepts in algebra: A technological approach.* Dedham, MA: Janson Publications, Inc. (republished in 1999 version, Chicago: Everyday Learning Corporation).

Fey, J. T., & Heid, M. K., with Good, R. A., Sheets, C., Blume, G. W., & Zbiek, R. M. (1991). *Computer-Intensive Algebra.* College Park, MD: University of Maryland.

Foletta, G. M. (2002, April). *High School Teacher and Her Students' Use of Representations while Using the CAS-Intensive Mathematics Curriculum.* Paper presented at the Research Presession to the 2002 Annual Meeting of the National Council of Teachers of Mathematics. Paper retrieved May 1, 2008, from www.nku.edu/~foletta/

Freda, A. (2008). To CAS or not to CAS. *Mathematics Teacher, 102*(1), 8-9.

Guin, D., Ruthven, K., & Trouche, L. (Eds.). (2005). *The didactical challenge of symbolic calculators: Turning a computational device into a mathematical instrument.* New York: Springer.

Heid, M. K. (1983). Calculus with muMath: Implications for curriculum reform. *The Computing Teacher, 11,* 46-49.

Heid, M. K. (1987). *Algebra with Computers in XX High School: A description and an evaluation of student performance and attitudes* (Report submitted to XX Area Schools Board of Education). Unpublished manuscript.

Heid, M. K. (1988). Resequencing skills and concepts in applied calculus through the use of the computer as a tool. *Journal for Research in Mathematics Education, 19,* 2–25.

Heid, M. K. (1996). A technology-intensive functional approach to the emergence of algebraic thinking. In C. Kieran, N. Bednarz, & L. Lee (Eds.), *Approaches to algebra: Perspectives for research and teaching* (pp. 239–255). Dordrecht: Kluwer.

Heid, M. K. (2002). CAS in secondary mathematics classes—The time to act is now! *Mathematics Teacher, 95*(9), 662–667.

Heid, M. K. (2003). Theories That Inform the Use of CAS in the Teaching and Learning of Mathematics. In J. T. Fey, A. Cuoco, C. Kieran, L. McMullin, & R. M. Zbiek (Eds.), *CAS in mathematics education* (pp. 33–52). Reston, VA: NCTM.

Heid, M. K., & Blume, G. W. (Eds.). (2008a). *Research on technology in the learning and teaching of mathematics: Vol. 1, Research Syntheses.* Charlotte, NC: Information Age.

Heid, M. K., & Blume, G. W. (2008b) Algebra and functions. In M. K. Heid & G. W. Blume (Eds.), *Research on technology in the learning and teaching of mathematics: Vol. 1, Research Syntheses* (pp. 55–108). Charlotte, NC: Information Age.

Heid, M. K., Blume, G., Hollebrands, K., & Piez, C. (2002, April). *The Development of a Mathematics Task Coding Instrument (MaTCI).* Paper presented at the Research Presession to the 2002 Annual Meeting of the National Council of Teachers of Mathematics. Paper retrieved May 1, 2008, from www.nku.edu/~foletta/

Heid, M. K., Blume, G., Zbiek, R. M., & Edwards, B. (1998/1999). Factors that influence teachers learning to do interviews to understand students' mathematical understandings. *Educational Studies in Mathematics, 37*(3), 223–249.

Heid, M. K., Hollebrands, K., & Iseri, L. (2002). Reasoning, justification, and proof, with examples from technological environments. *Mathematics Teacher, 95*(3), 210–216.

Heid, M. K., & Kunkle, D. (1988). Computer-generated tables: Tools for concept development in elementary algebra. In Arthur F. Coxford (Ed.), *The ideas of algebra: K–12, 1988 Yearbook of the National Council of Teachers of Mathematics* (pp. 170–177). Reston, VA: NCTM.

Heid, M. K., Zbiek, R. M., & Blume, G. (1997). Empowering mathematics teachers in computer-intensive environments. In J. Harvey (Ed.), *Models for mathematics technology teacher development programs, MAA Reports No. 3* (pp. 79-93). Washington, DC: Mathematical Association of America.

Heid, M. K., Zbiek, R. M., Blume, G. W., & Choate, J. (2004). Technology-intensive mathematics [CD]. Unpublished materials.

Heid, M. K.(1985). An exploratory study to examine the effects of resequencing skills and concepts in an applied calculus curriculum through the use of the microcomputer (Doctoral dissertation, University of Maryland College Park, 1984). *Dissertation Abstracts International, 46A,* 1548.

Iseri, L. (2004). Algebra students' developing symbolic reasoning in the context of a computer algebra system (Doctoral dissertation, The Pennsylvania State University, 2003). *Dissertation Abstracts International, 64,* 4397.

Kendal, M., & Stacey, K. (2001). The impact of teacher privileging on learning differentiation with technology. *International Journal of Computers for Mathematical Learning, 6*(2), 143–165.

Lagrange, J.-B. (1999). Complex calculators in the classroom: Theoretical and practical reflections on teaching pre-calculus. *International Journal of Computers for Mathematical Learning, 4*(1), 51-81.

Mathematics Educators Exploring Computer Algebra Systems (MEECAS). *Welcome to the MEECAS Website.* Retrieved May 1, 2008, from meecas.org

Monaghan, J. (1997). Teaching and learning in a computer algebra environment: Some issues relevant to sixth-form teachers in the 1990s. *The International Journal of Computer Algebra in Mathematics Education, 4*(3), 207–220.

Pierce, R., & Stacey, K. (2004). A framework for monitoring progress and planning teaching towards the effective use of computer algebra systems. *International Journal of Computers for Mathematical Learning, 9*(1), 59–93.

Trouche, L. (2004). Managing the complexity of human/machine interactions in computerized learning environments: Guiding students' command process through instrumental orchestrations. *International Journal of Computers for Mathematical Learning, 9*(3), 281–307.

Zbiek, R. M. (2002a, April*). Perspectives on classroom-based research on the teaching and learning of mathematics in the context of technology: Overview of content and goal.* Paper presented at the Research Presession to the 2002 Annual Meeting of the National Council of Teachers of Mathematics. Paper retrieved May 1, 2008, from www.nku.edu/~foletta/

Zbiek, R. M. (2002b, April). *MAGICAL framework describing the nature of students' use of representations.* Paper presented at the Research Presession to the 2002 Annual Meeting of the National Council of Teachers of Mathematics. Paper retrieved May 1, 2008, from www.nku.edu/~foletta/

Zbiek, R. M., & Finken, T. M. (2002, April). *Students' complex and nested uses of shared representations: An example with linear functions in a technological environment.* Paper presented at the Research Presession to the 2002 Annual Meeting of the National Council of Teachers of Mathematics. Paper retrieved May 1, 2008, from www.nku.edu/~foletta/

Zbiek, R. M. & Heid, M. K. (2009). Using computer algebra systems to develop big ideas in mathematics. *Mathematics Teacher, 107*(7), 540–544.

PART III

3-D GEOMETRY

CHAPTER 10

THREE-DIMENSIONAL CITIZENS DO NOT DESERVE A FLATLANDERS' EDUCATION

Curriculum and 3-D Geometry[1]

Claudi Alsina

INTRODUCTION

Our chief concern in this paper is to contribute to a wider consideration of 3-D Geometry in the compulsory mathematical curriculum. As a result of research results and practical experiences, space geometry in the curricular framework has experienced positive growth. In the old days the special knowledge was restricted to a few figures (polyhedra-spheres-cylinders-cones) and transformations, and no attention was given to the essential aim of the subject: to develop visual thinking and mathematical reasoning. Nevertheless, beyond the curricular frame, the real time dedicated to geometry in the classroom is still low and the level is weak. In general, plane geometry wins over space geometry. But many research studies have shown

[1] To Karl Menger and Marshall H. Stone, in Memoriam

Future Curricular Trends in School Algebra and Geometry: Proceedings of a Conference. pages 147–154

an unacceptably low level of achievement in 2-D geometry. The 3-D case is even worse. Bearing this situation in mind, we will defend the interests of space geometry, and how it can be developed by integrating appropriate resources (from hands-on materials to technological devices). We will make reference to recent research of Anton Aubanell that we have been tutoring and, hopefully, we will clarify some key issues concerning this important topic.

WHAT IS REALITY?

$\mathfrak{R}^3$ is not our reality. The real world is our house and our land; it is the objects we use and the transportation we ride, it is our social and cultural life... reality is a complex issue and merits a mathematical approach.

The real world is not the recreational problems paradigm of crossing trains, sheep in boats crossing rivers, castles with lakes, etc. Reality is not the Euclidean description of shapes. Many times we make the mistake of identifying reality with its mathematical models. Real space is not $\mathfrak{R}^3$: it is much more!

Moreover, our reality is not only a historical heritage. Our reality is also the 21st century, i.e., the problems and possibilities that we face today, from the use of new technological tools to environmental open problems.

We are three-dimensional people in a complex three-dimensional world,... and we have mathematics to discover and create action in this 3-D setting.

TOWARDS A SPATIAL CULTURE

The first aim for including 3-D geometry in the curricula is to give people opportunities for developing a spatial culture (Alsina, 2000a, b). We have identified eight principles that could guide our approach to 3-D geometry in the classrooms:

1. A first aim in learning 3-D geometry is to develop visual thinking, the basic competency for a spatial culture;
2. Common sense in 3-D needs to be developed since it is not genetic;
3. There is no need to follow the artificial ordering 1-D —2-D—3-D, since one can work at the same time in the different domains;
4. We can base our approach to 3-D geometry on interesting applied problems related to our reality;
5. We can mix different representations, models, and technologies to treat 3-D problems, since the multi-modeling gives richer results;
6. Spatial culture can facilitate connections and interdisciplinary approaches (history, art, environment and so on);

7. Spatial culture can promote the research spirit of our students and the discovery of intriguing results;
8. The final goal of a spatial culture is to develop human creativity.

A THEORETICAL FRAMEWORK

As a result of the research carried out in the last decades, many theoretical frameworks of reference have been developed, from the topological approach of Piaget and Inhelder (1967) to the latest social constructivism. All these theories, which may orient teaching and learning processes in mathematics, mark the importance of geometry.

Following Hershkowitz, Parzysz, and Dormoleu (1997) we may consider that geometry is seen by many students as the "science of the physical space and its mathematization" and that *visualization* is the key tool to facilitate geometry's learning. But visualization includes a rich combination of figures, diagrams, transformations, mental images, interactive figures, etc. (see Bishop, 1989; Del Grande, 1990). The theories of Internal Representation have clarified as well the role of images in the learning progression.

My personal point of view is that Freudenthal's ideas concerning "*mathematization*" are very relevant for space geometry. In this work, I follow the realistic mathematics framework, which focuses on *modeling and applications* (Malkevitch, 1998) for working 3-D geometry into the classrooms. Results from PISA show the need to devote more attention to this subject.

In the 1950s, under the direct influence of Hans Freudenthal, Pierre Marie and Dina Van Hiele started to develop a model concerning mathematical reasoning and its levels (Van Hiele, 1957). Since then, a lot of research has proven that the Van Hieles' model may be useful in geometrical thinking (Clements & Battista, 1992; Jaime, 1993; Gutiérrez, 2007). Let us recall that Van Hiele distinguishes 5 levels of reasoning, from the very basic simple observations to descriptive understanding/analytical thinking and more abstract settings including relations and proofs. Experimental studies (e.g., Clements & Battista, 1992) have shown that all students need to develop these different levels.

This model had a direct influence on the *Principles and Standards for School Mathematics* (National Council of Teachers of Mathematics, 2000) where the goal to attain at least the first three levels at grade K–12 guides the proposal of curricular development.

ON TEACHING RESOURCES AND TECHNOLOGIES

Since Maria Montessori's time, when small cubes and colorful prisms became a key tool for the learning of arithmetic, other resources have been produced for primary schools. The exhibition area at NCTM 2007 in Atlanta showed a very large collection of hands-on materials for school children.

The bad news is that there is a low production of hands-on materials for the higher grades where the commercial interest focuses on graphic calculators, software, polyhedra, etc. Nevertheless, as noted in Clements and Batista, 1992, the student's representation of space is constructed from active manipulation.

But the good news is that for working with 3-D geometry, one has at hand (and free of charge!), our own 3-D world, i.e., instruments for measuring, geometrical objects with clever designs, all kinds of technical devices, etc. In addition, one can easily produce materials which will facilitate teaching and learning activities (transparent polyhedra, soap films, mirrors, scale models, etc.).

The increasing power of technologies will have also a big impact on treating 3-D geometry. In the initial approach to 3-D geometry (at any level!) *virtual images can't substitute for the real observation and manipulation of hands-on materials.* To draw figures with ruler and compass by hand is not the equivalent of manipulating an applet making such drawings; the direct measure of a building can't be substituted by a virtual measure using Google Earth; the effective construction of an icosahedron with cardboard or plastic pieces can't be substituted by clicking the mouse to move a virtual polyhedron on the screen.

But technology has its own interest. It is a motivating tool and opens great learning opportunities. So there is no doubt that technology needs to be integrated in the process of mastering 3-D geometry. All drawing programs (Cabri-II, Geometry sketchpad, Geogebra, etc.), the new 3-D software (Cabri 3D, CAD, etc.), the interactive web sides (see Internet resources at the end), etc., may be excellent tools to use properly.

SOME NEW RESEARCH RESULTS OF ANTON AUBANELL

For many years Anton Aubanell (Aubanell, 2006) has been researching the subject of teaching 12–18 year-old students with hands-on materials. In 2005–2006, he had a one-year leave to develop, under my supervision, a research program whose goal was to answer the question: can all parts of the curriculum be treated by appropriate hands-on materials? The answer is yes. Results are available on the web site of our Department of Education (http://www.xtec.cat/aaubanel/). Data accumulated after many years of case studies in the IES Sa Palomera (Aubanell, 2006) showed that students become highly motivated when taking an active role in their own learning progress, when working cooperatively and developing a more visual approach.

In the case of 3-D geometry (Aubanell's study covered all parts of the curriculum) we have identified twelve uses of hands-on materials which are of interest for teaching and learning purposes. Figure 1 sums up our considerations.

FIGURE 1.

At the same time, we have seen that it is necessary to combine all kinds of representations and resources if we want to face rich tasks. The idea is to break the usual linear development of the textbook strategy and combine problem-solving, projects, technology, experimentation, etc.

IMPLICATIONS FOR THINKING, REASONING AND PROVING

Let's note that, beyond the concrete knowledge of shapes and transformations in space, the curricular work of 3-D geometry needs to help students develop visual thinking (with a rich variety of representation), improve their ways of mathematical reasoning (induction, analogy, generalization, etc.) and discover the need of proving. Greater attention to proofs and proving is a hot topic in today's education community (NCTM, 2000). Many researchers and national curricula recommend that the concept of *proof* and the activity of *proving* become an important goal for students' mathematical experiences throughout the grades. We are convinced, following ideas extensively developed by Gila Hanna (1996), that the *proofs* that are of interest for teaching are those that facilitate a better understanding of the results or concepts being faced. For making possible visual proofs we have developed (in Alsina and Nelsen, 2006) a full methodology on how to create what we call "images for understanding", where 2-D and 3-D images are combined.

GOOD CURRICULA DO NOT MEAN BETTER TEACHING

While we are now more optimistic than in the past about the role of space geometry in many curricula, it is unclear to what extent this has been really developed in classrooms. To make this possible, much more training on space geometry must be offered in pre-service and in-service teacher's courses and seminars.

Our own experiences in Spain and Argentina confirm also that the lack of geometrical knowledge among teachers can be compensated with attractive in-service actions (e.g., with geometry workshops) and actions in professional development.

TO SUM UP: GIVE 3-D GEOMETRY AN OPPORTUNITY

We are convinced that geometry merits further attention in the curriculum. But we have also seen that there is a lot to do with teachers before some progress can be appreciated. Often, teachers are not confident with the mathematics of 3-D geometry and the minimal-effort principle (textbook and blackboard) has a strong influence.

Our future citizens deserve our efforts towards a better learning of 3-D geometry.

REFERENCES

Alsina, C. (2000a). *Sorpresas geométricas.* Buenos Aires: OMA.

Alsina, C. (2000b). Gaudi's ideas for your classroom: Geometry for Three-Dimensional Citizens, *Selected Lectures ICME-9.* [CD ROM]. Tokyo: Makuhari-Tokyo.

Alsina, C. (2001). *Geometría y realidad en Aspectos didácticos de Matemáticas.* 8 ICE. (pp. 11–32). Zaragoza: Universidad de Zaragoza.

Alsina, C. (2007). Less chalk, less words, less symbols...more objects, more context, more actions. In W. Blum, P. L. Galbraith, H. W. Henn & M. Niss (Eds.), *The 14th ICMI study: Modeling and applications in mathematics education* (pp. 35–55). Berlin: Springer.

Alsina, C. (2007). Invitación a la tridimensionalidad. In P. Flores, F. Ruiz & M. Fuente (Eds.), *Geometría para el siglo XXI* (pp. 119–139). Badajoz, Spain: FESPM

Alsina, C., Burgués, C., & Fortuny, J. M. (1990). *Materiales para construir la Geometría.* Madrid: Síntesis.

Alsina, C., Fortuny, J. M., & Pérez, R. (1997). *¿Por qué Geometría? Propuestas didácticas para la ESO.* Madrid: Síntesis.

Alsina, C. & Nelson, R. B. (2006). *Math made visual: Creating images for understanding mathematics.* Washington, D.C.: Mathematical Association of America.

Aubanell, A. (2006). *Recursos materials i activitats experimentals en l'educació matemàtica a adecundària.* Barcelona: Dep. Educació, Generalitat de Catalunya.

Battista, M. T. & Clements, D. H. (1996). Students' understanding of three-dimensional rectangular arrays of cubes. *Journal for Research in Mathematics Education, 27*(3), 258–292.

Bishop, A. J. (1989). Review of research on visualization in mathematics education. *Focus on Learning Problems in Mathematics, 11*(1), 7–16.

Bolt, B. (1991). *Mathematics meets technology.* Cambridge, England: Cambridge University Press.

Clements, D. H., Battista, M. T. (1992). Geometry and spatial reasoning. In D. A. Grouws (Ed.), *Handbook of research on mathematics teaching and learning* (pp. 420–464). New York: MacMillan.

Corberán, R., Gutiérrez, A. et al. (1994). *Diseño y evaluación de una propuesta curricular de aprendizaje de la Geometría en Enseñanza Secundaria basada en el Modelo de Razonamiento de Van Hiele* (colección "Investigación" n. 95). Madrid: CIDE-MEC.

Del Grande, J. (1990). Spatial sense. *Arithmetic Teacher, 37*(6), 14–20.

Flores, P., Ruiz, F., & Fuente, M. (Eds.). (2006). *Geometría para el siglo XXI.* Badajoz, Spain: FESPM.

Guillén, G. (1997). *El modelo de Van Hiele aplicado a la geometría de los sólidos. Observación de procesos de aprendizaje.* Valencia: Universidad de Valencia.

Guillén, G. (2000). Sobre el aprendizaje de conceptos geométricos relativos a los sólidos: Ideas erróneas. *Enseñanza de las Ciencias, 18*(1), 35–53.

Gutiérrez, A. (1996). Visualization in 3 dimensional geometry: In search of a framework. In L. Puig & A. Gutiérrez (Eds.), *Proceedings of the 20th Conference of the International group for the Psychology of Mathematics Education* (Vol. 1, pp. 3–19). Valencia: Universidad de Valencia.

Gutiérrez, A. (1998). Las representaciones planas de cuerpos 3-dimensionales en la enseñanza de la geometría espacial. *Revista EMA, 3*(3), 193-220.

Gutiérrez, A., Jaime, A., & Fortuny, J. M. (1991). An alternative paradigm to evaluate the acquisition of the Van Hiele levels. *Journal for Research in Mathematics Education, 22*(3), 237–251.

Guzmán, M. de. (1991). *Para pensar mejor.* Barcelona: Labor.

Hanna, G. (1996). The ongoing value of proof. In L. Puig & A. Gutierrez. (Eds.), *Proceedings of the 20th Conference of the international group for the Psychology of Mathematics Education,* (Vol. 1, pp. 21–34). Valencia, Spain: Universidad de Valencia.

Hershkowitz, R., Parzysz, B., & van Dormolen, J. (1997) Shape and space. In A. J. Bishop, K. Clements, C. Keitel, J. Kilpatrick & C. Laborde (Eds.), *International Handbook of Mathematics Education.* (pp. 161–204). Dordrecht: Kluwer.

Jaime, A. (1993). *Aportaciones a la interpretación y aplicación del modelo de Van Hiele: La enseñanza de las isometrías del plano, la evaluación del nivel de razonamiento.* Valencia: Universidad de Valencia.

Lakatos, I. (1978). *Pruebas y refutaciones: La lógica del descubrimeinto matemático.* Madrid: Alianza.

Malkevitch, J. (1998). Finding room in the curriculum for recent geometry. In C. Mammana & V. Villani (Eds.), *Perspectives on the teaching of geometry for the 21st century: An ICMI study* (Vol. 5, pp. 18–24). Dordrecht: Kluwer.

Malkevitch, J. (1998). Geometry and reality. In C. Mammana & V. Villani (Eds.), *Perspectives on the teaching of geometry for the 21st century: An ICMI study* (Vol. 5, pp. 85–99). Dordrecht: Kluwer.

Malkevitch, J. (1992). *Geometry's future.* Lexington: COMAP.

Mammana, C. & Villani, V. (Eds.). (1998). *Perspective on the Teaching of Geometry for the 21st Century: An ICMI study* (vol. 5). Dordrecht: Kluwer.

National Council of Teachers of Mathematics. (2000). *Principles and standards for school mathematics.* Reston, VA: Author.

Plugh, G. (1976). *Polyhedra, a visual approach.* Londres: University of California Press.

Steen, L. A. (1994). *For all practical purposes.* Lexington: COMAP. New York: W.H. Freeman Co. Versión española: (1999). *Matemáticas en la vida cotidiana.* Madrid: Addison-Wesley.

Van Hiele, P. M. (1957). *El problema de la compresión en conexión con la comprensión de los escolares en el aprendizaje de la geometría [De problematiek van net Inzicht gedemonstreed wan het Inzicht von Schoolkindren in meetkundeleerstof].* Doctoral dissertation. Utrecht, Holanda: Univ. Utrecht.

Van Hiele, P. M. (1986). *Structure and insight: A theory of mathematics education.* London: Academic Press.

Veloso, E. (1998). *Geometria: Temas actuais e materais para professores.* Lisboa, Portugal: Instituto de Inovação Educacional.

INTERACTIVE RESOURCES ON THE INTERNET

http://www.nctm.org

http://www.fi.uu.nl

http://www.comap.com

http://www.ies.co.jp

http://www.cut-the-knot.com

http://www.uib.no

http://www-groups.des.st-and.ac.uk

http://members.xoom.com

http://www.li.net

http://www.geocities.com

http://www.profes.net

http://www.museo.unimo.it

http://www.ntu.edu.sg

http://www.upc.edu/ea-smi/personal/claudi/index.html

CHAPTER 11

MANIPULATING 3D OBJECTS IN A COMPUTER ENVIRONMENT

Jean-Marie Laborde

THE DIRECT MANIPULATION REVOLUTION IN COMPUTERS

During the last 30 years, new paradigms have developed due to the massive use of computers in all domains of human activities. Here the main profound change has been the "triumph" of computer use, based on direct manipulation design. This concept originated from the Research Lab at Rank Xerox in Palo Alto (1979) (Smith et al. 1982) and has been later popularized by Apple Computers (Lisa, Macintosh) and eventually available on PCs (Windows, PS2, Windows xx) and Unix based systems.

In the meantime, various pieces of software and/or computer environments have been developed with education in mind. One of the most famous and emblematic examples is probably LOGO [Papert].

Somehow later the concept of microworld (embedded precisely in LOGO) developed in various systems offering to their users the possibility to act more or less directly on the representations of mathematical objects. One privileged domain in mathematics has been geometry because of the clear relationship between geometric abstract entities and their "natural"

Future Curricular Trends in School Algebra and Geometry: Proceedings of a Conference. pages 155–167

representations on a computer screen. This is how software like Cabri (Laborde et al., 1990), Geometer's Sketchpad (Jackiw, 1991) or more recently Cinderella (Kortenkampf et al., 1999) appeared. (Many others have appeared, some of them cloning in an unethical way, more or less visibly, the above ones.) Nevertheless, for a long period of time, 3D standard geometry in a computer environment has been only accessible in reconstructing "by hand" perspective views of 3D objects using the available, essentially 2D, environments. What makes authentic 3D direct manipulation a specific case?

3D GEOMETRY AS A SPECIAL CASE

I like to quote some 9th-grade students in France: "Geometry? It is difficult, we are asked to put down so many things. We are asked to do proofs, it is complicated. One is supposed to behave as if things would not be known, but they are there, visible, on the diagram!" Further, when asked about solid geometry they continue with: "The problem is that we do not understand anything from the diagram. And it is proven that girls are less visual thinkers than boys..." Their teacher then concludes: "Our everyday life is in 3D space. We are faced continuously with 3D representations: Paintings, pictures, Assembly guides, Technical drawings,... Why is it so difficult for students to master Solid Geometry?"

This raises a number of questions about the role of diagrams and representations, about the discrepancy about our need to deal within a 3D world and its actual "understanding" as an abstraction, something needed to perform nonroutine tasks (Laborde, 2008).

Below is a possible ordinary standard representation of two lines lying in space and the question is "Do they intersect?"

Using a 3D environment the diagrams might be less stereotypic, possibly more realistic and moreover manipulable with the mouse: the students could move the scene and understand that some depth intervenes.

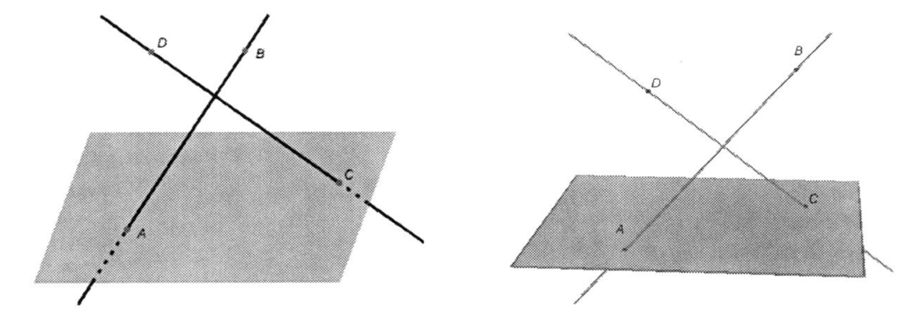

FIGURE 1. Do lines AB and CD intersect?

So 3D appears as a crucial place because our everyday world is a 3D world, because we are dealing with many 2D representations of this world (at practical levels, with pictures, paintings, diagrams) of objects of this 3-dimensional world, yet students have a lot of difficulties in "understanding" the geometry of those representations, and, as a matter of fact, because very often teachers have the feeling not to be equipped to teach the students this domain. All of this creates the need and the demand to help students and teachers, and since we have evidence that Dynamic Geometry Systems running on computers or on calculators (Cabri, GSP) have been very efficient in augmenting the quality of mathematics teaching/learning in general and not just for geometry), it is natural to conceive similar systems explicitly oriented to help students in mastering 3D representations.

In the Cabri project, we been paying attention to the possibility of developing such a system (today called Cabri 3D) since 1995, and we have had a first prototype since the late 1990s.

SOME ISSUES OF GENERAL NATURE ABOUT 3D ENVIRONMENTS

What have been the issues encountered in considering the idea of a computer environment for 3D geometry and 3D representations and what design decision have made to create our current Cabri 3D?

- How to represent 3D objects on a 2D flat surface. The question of the viewer's perspective;
- How to manipulate 3D objects having only essentially a 2D pointing device at disposal?
- How to avoid the "white sheet" effect?
- How to deal with "unrealistic" representations?
- What is a meaningful shape of a plane?
- ...

It would be difficult to address here, just in one paper, all these issues. For this reason I go for a tour linking, from a cultural perspective, math and art, especially in 3D and 4D.

After the work of Alberti, Dürer, Desargues, and others, mankind has been able to achieve meaningful representations of spatial objects on a flat canvas. After the additional work of many talented researchers and engineers in the computer science research and computer industry, it has been possible to automate the actual flat represention of abstract 3D objects. Why not to follow the steps of Alberti in one more dimension and to obtain perspective constructions of 4-dimensional objects in our ordinary 3D world? This is what I will look at in the sequel.

AN INCURSION-EXCURSION INTO 4D

To explain the issue, we consider the idea of projecting (i.e. view in perspective) objects of the 4-dimensional space into our 3D world and then manipulate them on a computer screen.

FIGURE 2. "Corpus Hypercubicus", by Salvador Dali

Here, using current ordinary technology, the actual representation on the computer screen will be the result of a second projection from 3D to 2D. Nevertheless the actual direct manipulation of the 3D projection will involve 3D objects and user feedback will match an authentic 3D space manipulation, in essence different from more common 2D feedback offered by most of the 2D DGS.[1]

To illustrate this I would like to bridge a famous painting, Salvador Dali's "Corpus Hypercubicus"[2] with the 4D mathematical object that gave the painting its name, the hypercube, an object that is actually the cube of dimension four.

Knowing Dali's rapture for words in "...que" like in "hypercubique", "métaphysique", "paranoïaque"... this strange title "Corpus Hypercubicus" might be viewed simply as a provocative play on words. Looking closer one can nevertheless discover that the "cross" is made of some central cube extend along its six faces by six other cubes of the same side and an additional eighth cube placed at the bottom on this construction.

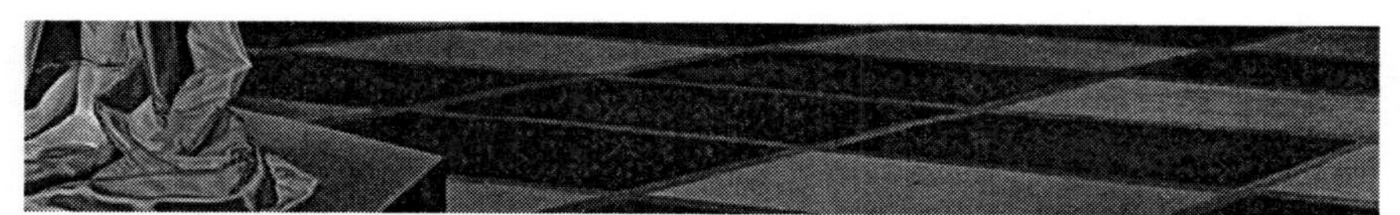

FIGURE 3. Six dark brown square tiles on the floor in Dali's painting.

On the ground one can also see (in perspective) in the tiling an interesting pattern: six squares arranged in the following way, when viewed from the top (Figure 4).

Here is the shape of a cross to be recognized as well the net of a cube well-known even by many elementary school mathematics students.

In addition, one immediately notes that the Dali's process to go from a cube to his "hyper-cross" is just the generalization of starting from a square, extending it along its four sides by other same sized squares and completed with an additional square here on the left and, in Dali's case, at the bottom. The natural question is then, can we think of Dali's construction as the (hyper)net of a hypercube? This would explain in a convincing way the semantic of the title "Corpus Hypercubicus" coined by Dali. (Note that the

[1]We have to be aware of more advanced technology allowing the user to perceive the representation in 3D, e.g. using special 3D glasses, bypassing to some extent the limitations of flat screens. To immerse the user more completely in 3D one needs then a replacement for the ordinary 2D mouse. If not, the results can be baffling to understand and frustrating to manipulate. Unfortunately such advanced authentic 3D space-pointing devices are not yet commercially available and then highly expensive.

[2]This impressive painting, which dates from 1954, is in the Metropolitan Museum of Art in New York.

FIGURE 4. Six tiles forming the cross-shaped net of a cube.

net of a cube is a plane or 2D object, and that the net of a 4D cube is just an ordinary spatial or 3D object).

We have to give sense to a cube in dimension 4 and then to see how we could "fold" the hyper-cross to obtain some 4-dimensional cube.

Apparently people commonly agree on the fact that following object is an accurate representation of a hypercube (Figure 5).

It is made of two 3D-cubes whose analog vertices are connected with additional edges. This generalizes the construction of a cube obtained from 2 2D-cubes (i.e. 2 squares) whose corresponding vertices are connected by 4 edges (Figure 6).[3]

[3]For the sake of consistency one can look at this in one dimension less: actually taking 2 1D-cubes (i.e. segments) and, connecting their vertices using 2 additional vertices, one obtains a 2D-cube (i.e. a square). In addition let us consider a line segment (a 1-cube), extend it with two copies of it along its boundary, i.e. its two endpoints. Adding an additional fourth line segment at one end, one obtains a chain of four line segments, actually the "linear" net of a square!

Below is a Cabri 3D construction showing the generalization process: From dimension 0 to 1 to obtain a line segment, from 1 to 2 to obtain a square, from 2 to 3 to obtain a cube (in thin, we have represented the "bottom" element starting to rotate in the folding process)

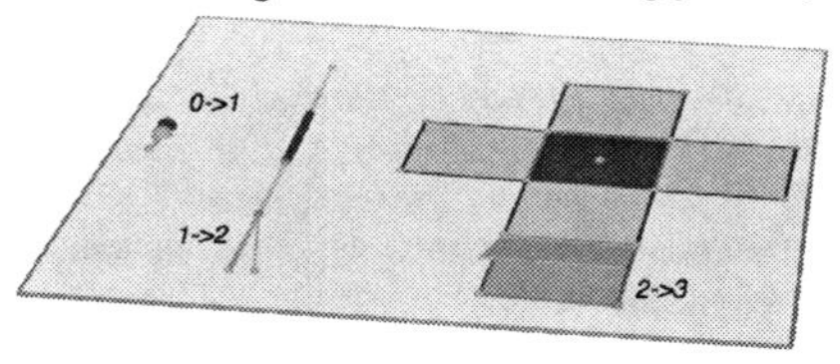

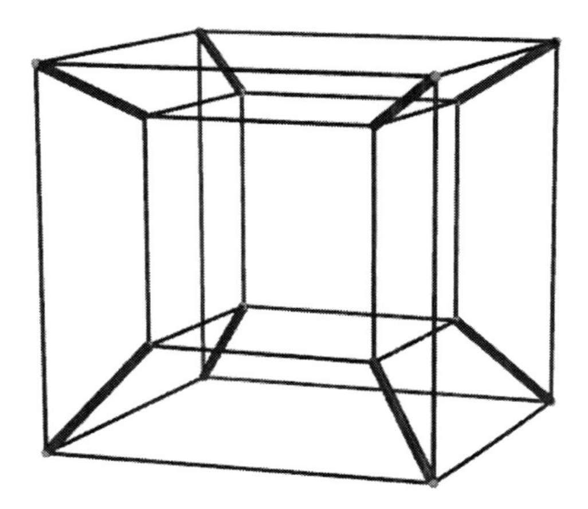

FIGURE 5. A hypercube represented by two cubes with corresponding vertices connected by 8 edges.

To complete the tour, we need to understand what will happen when we fold in the 4-dimensional space the hyper-cross, or actually what will be the 3D-perspective drawing of the net being folded, rotating 7 pieces of the 8 cubes, by 90°.

The best is to recall what a rotation is. In 2D or in the plane, it has

- a fixed point,
- and it is defined by an angle α in $[0,\pi]$.

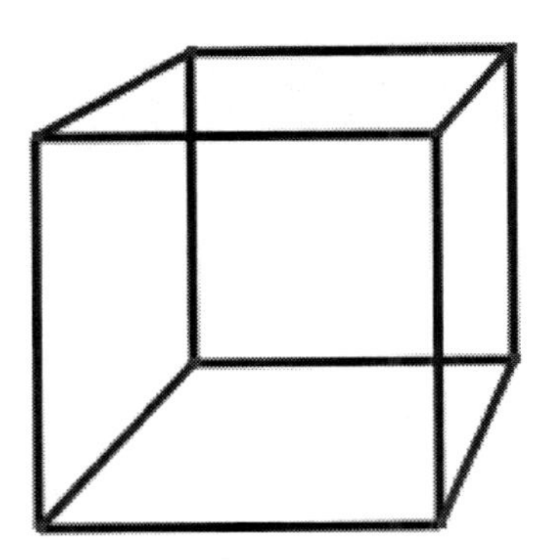

FIGURE 6. A cube obtained by connecting with edges corresponding vertices of two squares.

"Similarly" a rotation in the space is characterized by

- an oriented axis (a fixed line) and
- an angle α in $[0,\pi]$.

Both transformations preserve distance (they are isometries), have (at least) one fixed point, and preserve orientation.

So in n dimensions a rotation is defined as a direct isometry with (at least) one invariant point.

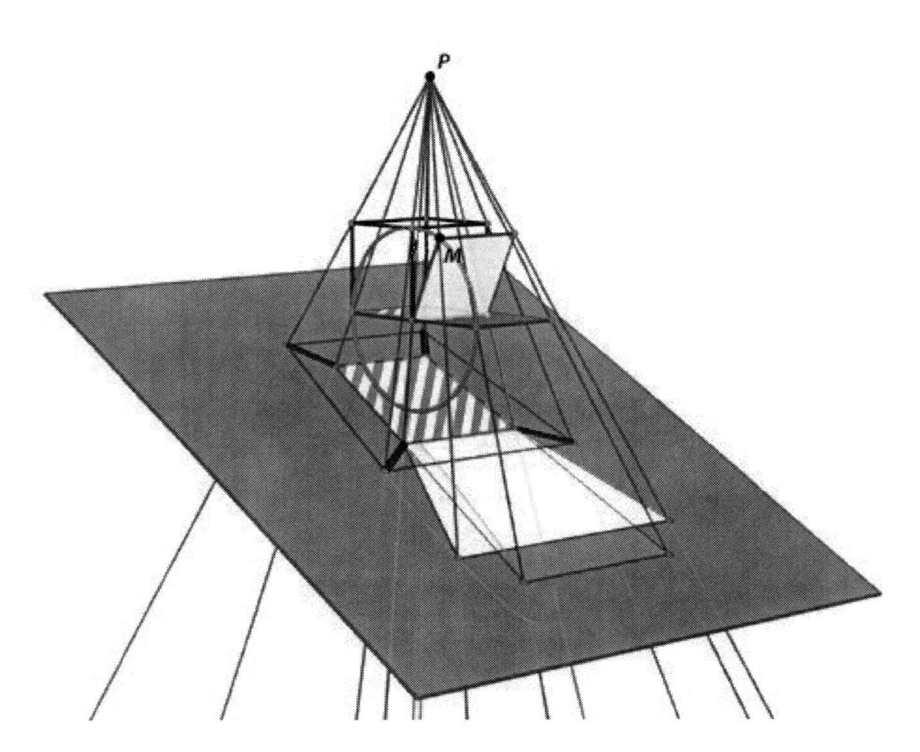

FIGURE 7. This is the construction to be "generalized" from 2D to 3D.
Consider a small "horizontal" white-gray hashed square part of the boundary of a cube.Next to it, sharing an edge as a hinge, is another square of same size (and gray). We rotate it to make it vertical in folding it toward a 3 dimensional cube (wireframe in black). The white square represents one of its intermediate positions. Through a perspective (central projection from P) we obtain one 2D image in an oblique gray plane. Moving point M allows to understand the successive polygonal shapes (in our case, for symmetry reasons, trapezoids) the white image will take from its starting shape in gray to its final in white-gray, when the square will "pass" 90° and perform a 180° rotation. Vertices of the image of the white square describe conics.

It can be shown (easily...) that in dimensions 2 and 3 corresponding transformations actually match the ordinary plane and spatial rotations. In 4D we have the following situation (less easy...):

For any rotation in $\mathbb{R}^4$, there exist 2 globally invariant orthogonal planes. Its restriction to these planes are 2D rotations of angle α and β in $[0,\pi]$, and any half-line is transformed into another half-line using an angle comprised between α and β.

FIGURE 8. 16 successive snapshots demonstrating the folding from the hyper-cross to the Great Arch (see below)

FIGURE 9. The Great Arch in Paris (photocredit: Clasohm, 2003)

In the special case $\alpha = 0$, it is possible to have one plane pointwise invariant and reduce the rotation to an ordinary 2D rotation in any orthogonal plane.

With some work, generalizing the way we can look—in perspective—to what happens to a square when it is, in 3D, rotated along the hinge constituted by one its edges (part of its boundary), one can "construct" in Cabri 3D the perspective dynamic representation of a cube rotated in 4D along the hinge made of one its six faces (part of the boundary):

Doing so what we reach progressively is exactly a hypercube.

It is interesting that the Great Arch erected in Paris for the 200th anniversary of the French Revolution is based on the same structure. In this building, four parallel squares of the original 4-cube have been removed, for the spectator to see "inside."

CONCLUDING WORDS

Connecting math and culture, we have been able to explore some 3D (and 4D) math from a cultural perspective. I am wondering, unfortunately, how

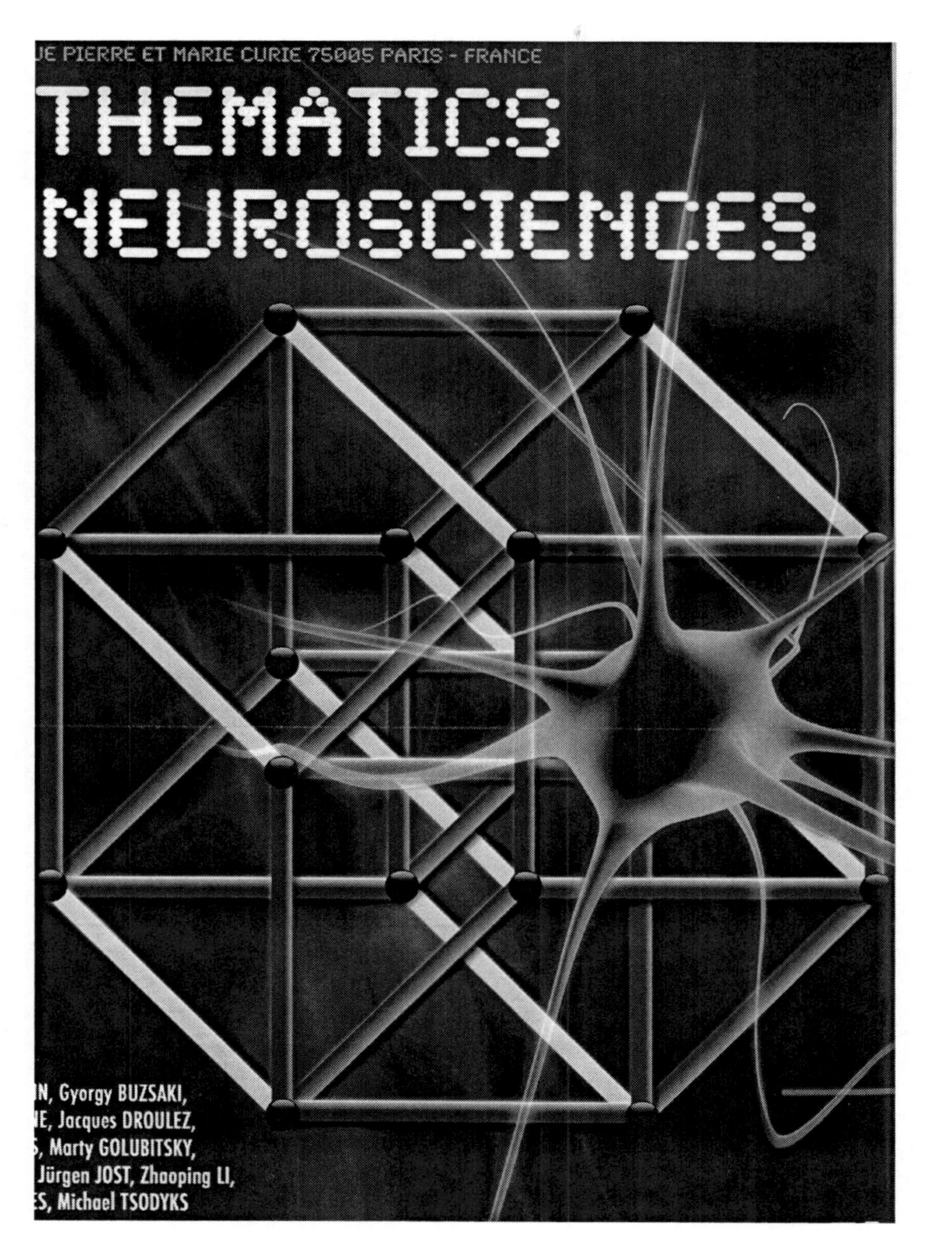

FIGURE 10. A 4-D cube inviting to a conference on neurosciences

wenn wir das Verfahren anwenden, das in den Abb. 153 bis 157 benutzt war. Die begrenzenden Polyeder des Zells werden durch ein

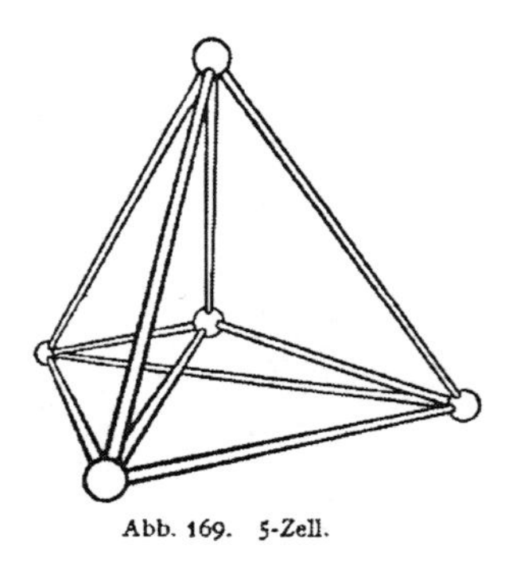

Abb. 169. 5-Zell.

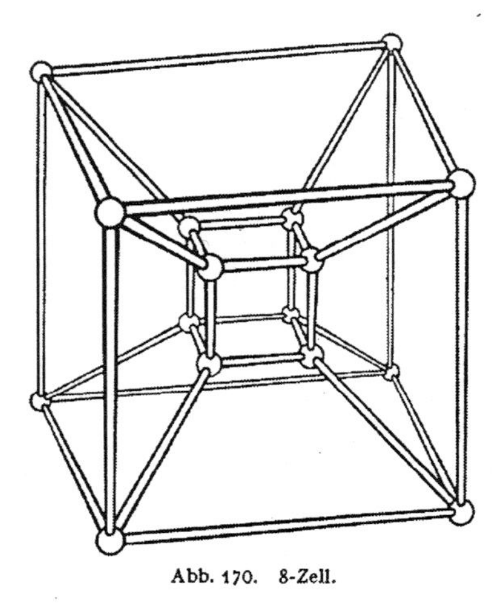

Abb. 170. 8-Zell.

FIGURE 11. 4-D tetrahedron and cube as representes in Hibert and Cohn-Vossen's book, Geometry and imagination.

many people looking at the Grande Arche have made the connection with a hypercube or even more with Dali's painting. Nevertheless, especially through effective use of modern technology, enhancing math, sciences and culture, learning and teaching will hopefully develop more widely in a global culture.

Hypercubes are becoming more and more popular. Their structure is intensively used in modern parallel computing. Below is a poster advertising a recent conference on another theme, neurosciences:

If hypercubes are so ubiquitous when did they first appear? Apparently nobody knows too much, even Thomas Banchoff, who is probably the first mathematician having contemplated the beauty of the connection between Dali's painting and the hypercube. Banchoff thinks that some monk in the middle age (in Portugal?) represented, even so "early", something very similar to Dali's hyper-cross. Yet another ancient source, even if its does not mention the word hypercube nor tesseract, can be found in the Diderot-D'Alembert Encyclopédie, in the article Dimension: "An intelligent person could look at the passing of time as a fourth dimension."

As concluding remark I would like to display the following representation of spatial geometric objects, designed under the supervision of David Hilbert (Figure 11).

One can surely trace here a resemblance with the images of 3D objects obtained using Cabri 3D. Actually this is not by coincidence because, back to the issues I raised above, the question of the "quality" of actual representation of mathematical objects has been shown to be crucial. Something that was also apparently clear to Hilbert when he decided to create representations exaggerating and even "deforming" or "reforming" the mathematical concept in display. I have to say that, like many mathematicians, I have been strongly impressed by Hilbert and Cohn-Vossen's book during my study and that part of the Cabri 3D design has been intensely marked by that wonderful book.

REFERENCES

Bainville, E., & Laborde, J.-M. (2004). *Cabri 3D* (Version 2.1.2) [computer software]. Grenoble: Cabrilog.

Banchoff, T. F. (1990). *Beyond the third dimension*: Geometry, computer graphics, and higher dimensions. New York: Scientific American Library.

Dali, S. (1954). *Corpus Hypercubicus* [Painting]. Metropolitan Museum of Art, New York.

Diderot, Denis, & Jean Le Rond d'Alembert. (1751) *Encyclopédie ou Dictionnaire raisonné des sciences, (Dimension) des arts et des métiers.* Paris.

Hilbert, D. & Cohn-Vossen, S. (1932). *Anschauliche Geometrie,* Springer Verlag p. 133, in English, *Geometry and the Imagination,* AMS Chelsea Publishing: New York 1952.

Jackiw, N. (1991). *The Geometer's Sketchpad* (Version 4) [computer software]. Berkeley: Key Curriculum Press.

Kortenkamp, U., & Richter-Gebert, J. (1999). *Cinderella* [computer software]. Berlin: Springer.

Laborde, C. (2008). Experiencing the multiple dimensions of mathematics with dynamic 3D geometry environments: Illustration with Cabri 3D. *The Electronic Journal of Mathematics and Technology, 2*(1), 38–53

Laborde, J.-M. et al. (1990). *Cabri II Plus* (Version 2) [computer software]. Grenoble: LSD2-IMAG-Cabrillog.

Laborde, J.-M. (2006). *Animation of the Corpus Hypercubicus.* Cabrilog, http://gallery.cabri.com/en/misc.html.

Papert, S. (1980). *Mindstorms: Children, computers, and powerful ideas.* New York: Basic Books Inc.

Smith, D. C., Irby, C., Kimball, R., Verplank, B., & Harslem, E. (1982). Designing the star user interface. *Byte, 74*(4), 242–282.

von Spreckelsen, J. O., & Reitzel, E. (1989). *La Grande Arche de la Fraternité,* Paris.

Wikipedia (2008). *SO*(4). Retrieved July 29, 2009, from http://en.wikipedia.org/wiki/SO(4).

CHAPTER 12

ALGEBRA AND GEOMETRY

From Two to Three Dimensions

Thomas F. Banchoff

Future curricular innovations in the relationship between algebra and geometry will feature color patterning and interactive computer graphics. This paper will explore two aspects of this relationship: geometric interpretations of algebraic formulas by means of decompositions of geometric figures in two and three dimensions, and counting activities involving two- and three-dimensional arrays. Illustrations in the text are static, and interactive versions are accessible at the following website: http://www.math.brown.edu/~banchoff/CSMCAlgebra-Geometry.html.

These innovations are particularly important in the topic of the parallel session from the CSMC International Conference, namely the extension of theorems in the second dimension to the third. A number of these topics were the subject of the author's recently published article in the 2008 NCTM Yearbook "Algebraic Thinking and Geometric Thinking" (Banchoff, 2008), and the presentation at the conference featured geometric decompositions to illustrate algebraic results such as $a^2 - b^2 = (a - b)(a + b)$, extended to $a^3 - b^3 = (a - b)(a^2 + ab + b^2)$. Rather than repeat the same examples in this report, it is even better to come up with a new example of a similar sort, directly related to other talks in the conference.

In his presentation on algebra at the CSMC International Conference, Al Cuoco from the EDC presented an example, namely an interpretation

Future Curricular Trends in School Algebra and Geometry: Proceedings of a Conference. pages 169–181

of the difference of two squares: $(a + b)^2 - (a - b)^2 = 4ab$. He showed that it is directly related to the inequality between the arithmetic mean and the geometric mean, namely:

$$\frac{a+b}{2} \geq \sqrt{ab}\,.$$

He suggested that there should be a geometric diagram parallel to the algebra, and there is.

The most symmetrical geometric interpretation, developed during that lecture, is contained in the following picture, with four rectangles with sides a and b, and an interior square with side $a - b$ (Figure 1).

The picture is different for $a > b$, $a = b$, and $a < b$, as shown in the following set of three pairs, where the pieces are placed together on the left-hand side and spread apart on the right-hand side (Figure 2).

Note that when $a = b$, the interior square disappears, while in the case where $a < b$, the interior square has negative side length. The square of a negative number is positive so all five pieces have positive area. Note also that when $a > b$, there are two ways of arranging the four rectangular pieces,

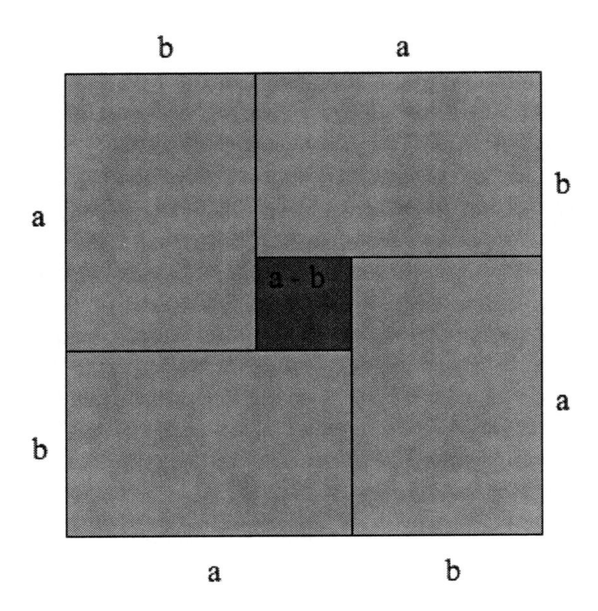

FIGURE 1.

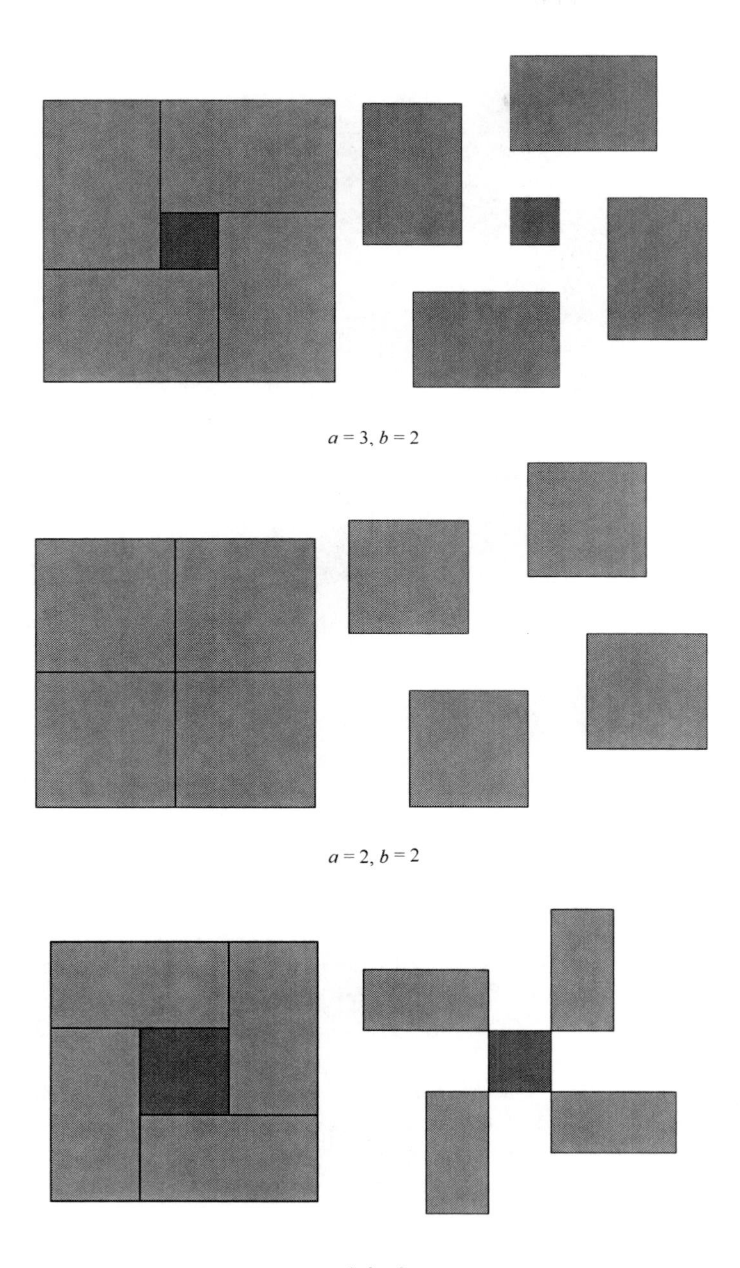

FIGURE 2.

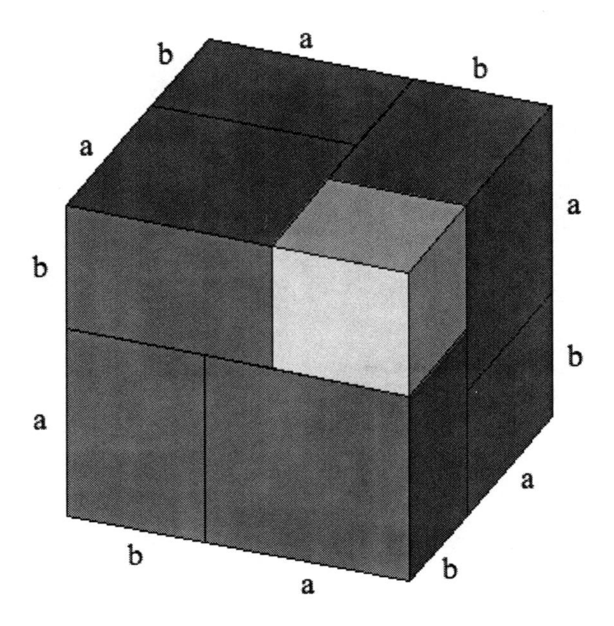

FIGURE 3.

mirror images of each other, and when $a < b$, the orientation is opposite from the orientation when $a > b$.

In the spirit of this conference, there is a built-in challenge, namely to find a 3-dimensional analogue of the 2-dimensional decomposition. The algebraic expression is $(a + b)^3 - (a - b)^3 = 6a^2b + 2b^3$. We can't expect a decomposition as symmetric as the one we found in the plane, but there is still a highly-symmetric decomposition of the region between two cubes, given by the following model. The two light gray cubes with side length b are at a pair of opposite corners of the dark gray inner cube with side length $a - b$, and all of the remaining gray pieces are prisms with height b and square bases with side length a. Once again, when $a = b$, the interior cube has zero side length. In the left-hand side of Figures 4 and 5, the pieces of the model are together and in the right-hand side, the pieces are spread apart.

The 3-dimensional case includes a surprise that does not appear in the plane, namely the case where $b > a$ so that the two light gray cubes overlap in a small inner cube with three negative sides and the total volume is negative. The sum of the (positive) volumes of the six dark gray prisms and the two light gray cubes is then equal to $(a+b)^3$ plus the absolute value of the volume of the inner cube.

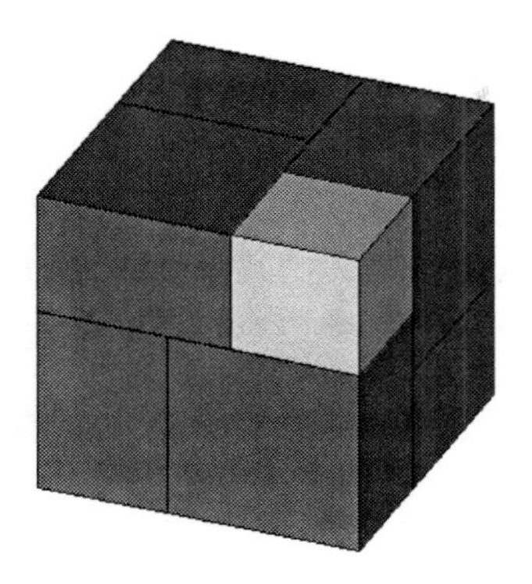

$a = 3$, $b = 2$ (the dark gray cube has edge length $a - b$)

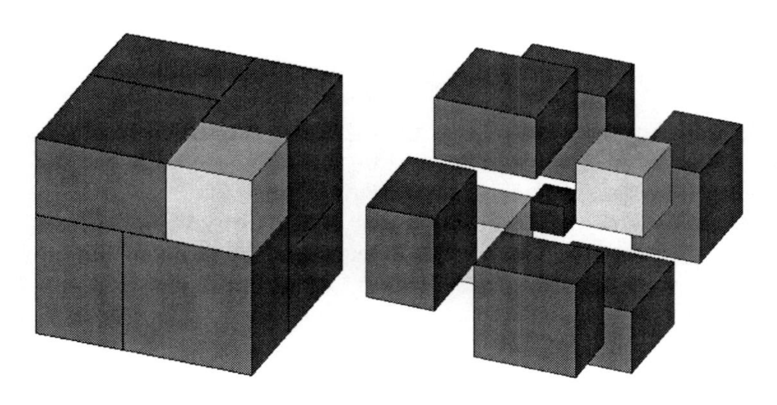

$a = 2$, $b = 2$

FIGURE 4.

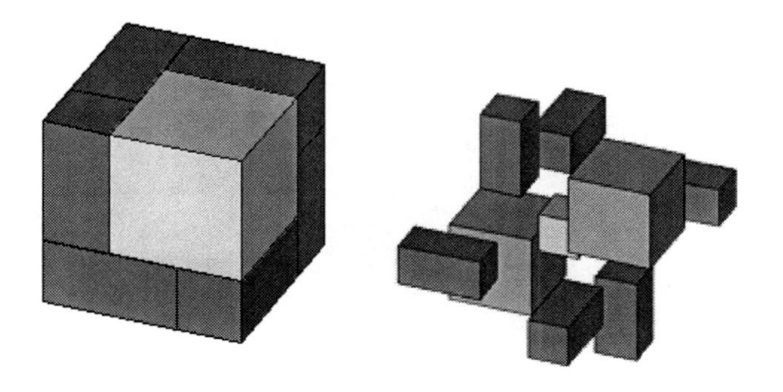

$a = 1$, $b = 2$ (the light gray cube has negative edge length $a - b$)

FIGURE 5.

Claudi Alsina, the commenter on the presentations in the second parallel session, has a number of similar decompositions in his book with Roger Nelson (Alsina and Nelson, 2006).

FROM THE COMMUTATIVE LAW IN THE PLANE TO THE ASSOCIATIVE LAW IN THREE-SPACE

In the two parallel sessions connected with going from 2- to 3-dimensional geometry, Michael Battista presented the results of some research indicating that young children sometimes have difficulty seeing a collection of blocks in a rectangle as an array of a certain number of rows with the same number of blocks in each row, or a certain number of columns with the same number in each column (Battista, 1998). A student who does see that pattern will recognize that a times b equals b times a for positive integers a and b. A student who doesn't recognize the array might easily miss the crucial algebraic point.

In a graphic representation, we can emphasize the structure of a rectangular $n \times m$ array of squares by separating it into n rows with m squares in each row or n squares in each of m columns (Figure 6).

We can do the same with a 3-dimensional collection of blocks illustrating the associative property. Usually this is expressed in terms of the equality of two expressions involving parentheses, for example $a \times (b \times c) = (a \times$

FIGURE 6.

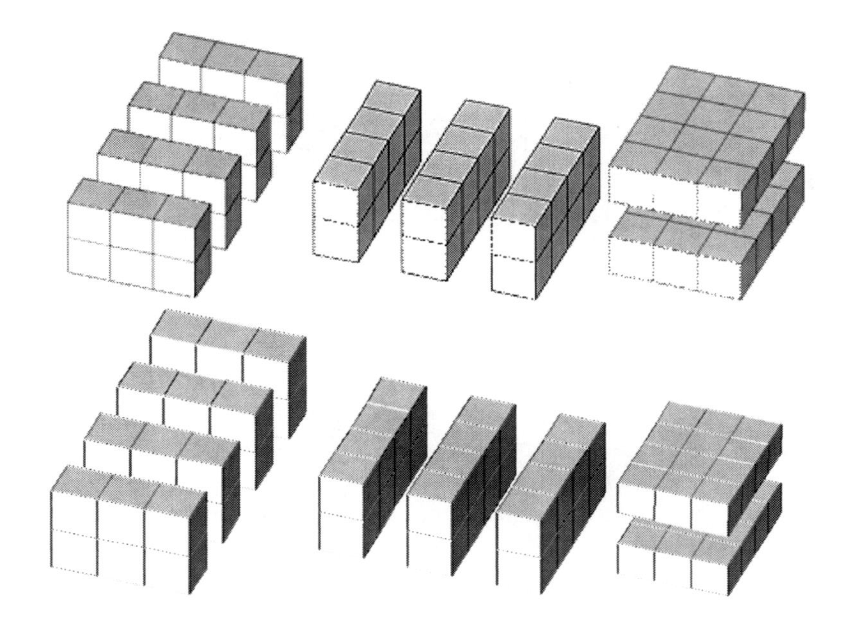

FIGURE 7.

$b) \times c$, but the geometric content of the statement is that is does not make any difference in which order we multiply three numbers, so $(a \times b) \times c = (b \times c) \times a = (c \times a) \times b$. Once again, we can emphasize the structure of a three-dimensional array of cubical blocks by spreading them apart in three different directions (Figure 7).

In these diagrams, the lines parallel to the *x*-axis are colored dark gray, those parallel to the *y*-axis are colored black, and those parallel to the *z*-axis are colored light gray. The rectangles parallel to the *xy*-plane are colored white, those parallel to the *yz*-plane are colored dark gray, and those parallel to the *xz*-plane are colored light gray.

If we remove the squares, then the black, dark gray, and light gray segments remain, and counting the number of colored segments is another kind of counting problem investigated by Michael Battista and his colleague (Figure 8).

"Flatland: the Movie" is another stimulus that leads to the same kinds of counting problems. In the parallel session, a short excerpt from this film introduced a topic for a geometry lesson that could take place in middle school or high school, involving counting the numbers of squares in various plane configurations. The precocious grandchild Hex outstrips her grandfather A Square by suggesting that patterns in the plane should have

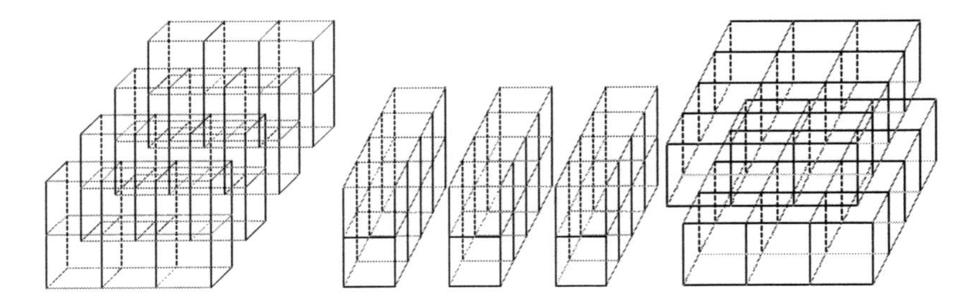

FIGURE 8.

analogues in the third dimension. In particular if a 3 by 3 square contains nine unit squares, then an analogous object in the next higher dimension will have 3 by 3 by 3 "super squares", or 27 units. Accompanying the film at the website http://www.flatlandthemovie.com, are a number of worksheets, one of which takes off from this question and leads to a number of exercises for individual study or for class assignments. The two- and three-dimensional portions of that worksheet appear below, as an example of ways such problems can be used in classrooms

HEX EXPLORES THE SUBDIVIDED SQUARES

"Grandfather," Hex asks, "What exactly is a dimension?"

When A Square begins to answer Hex, he starts with a single point, with zero dimensions, and moves it three units to trace out a segment subdivided into three unit segments (Figure 9).

The subdivided edge has four vertices and three unit segments.

He then moves the subdivided edge in a perpendicular direction, keeping it parallel to itself, to trace out a square region, subdivided into unit squares. Hex figures out the number of squares, namely nine.

Figure 10 is a 3×3 subdivided square.

In addition to counting the number of squares, Hex can learn more about the subdivided square by figuring out the number of vertices and the number of unit segments.

In a 3×3 subdivided square, there are:

FIGURE 9.

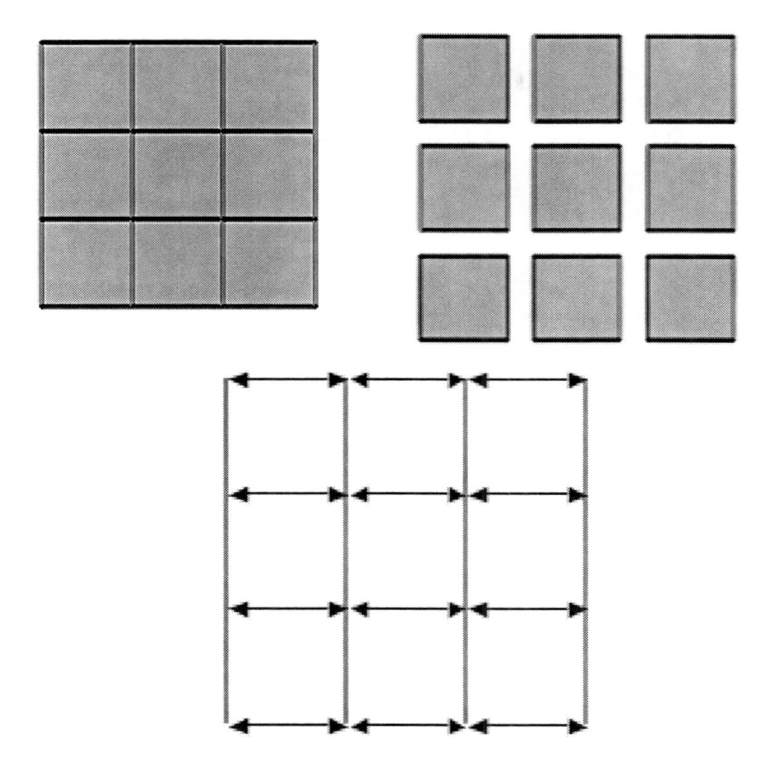

FIGURE 10.

- vertices
- horizontal unit segments
- vertical unit segments
- total number of unit segments
- unit squares

To make things clearer, A Square can work out a specific case, for example the 2×2 subdivided square, with each two unit segments on each side (Figure 11).

The number of vertices is V = 9, three on each of the three horizontal rows. The number of unit segments is E = 12, with 6 vertical segments and 6 horizontal segments. The number of unit squares is S = 4.

Now draw a 4×4 subdivided square. How many vertices does it have, and how many unit segments and how many squares?

Compile your answers for the 3×3 case and the 4×4 case in the following table. The first and second rows are already filled in:

FIGURE 11.

Number of subdivisions of each edge	Verticies (V)	Unit Segments (E)	Unit Squares (S)	V + E + S	V – E + S
1	4	4	1	9	1
2	9	12	4	25	1
3					
4					

What patterns do you see?

Can you predict what some of the numbers will be for a 5 x 5 subdivided square? Draw a diagram and use it to check your predictions.

ON TO THREE-DIMENSIONAL SPACE

Hex isn't satisfied just with finding patterns in the plane. She suggests to her grandfather that it is possible to go further and look for patterns in a space of three dimensions. What happens in our space of three dimensions? Moving a 3 x 3 subdivided square would produce a 3 x 3 x 3 subdivided cube. Even though Hex can't see it, she uses her imagination to predict that it will have 27 unit cubes (Figure 12).

We in Spaceland can see a subdivided cube and if we separate the cubes slightly, we can see what is inside. We can see the 27 unit cubes, in 3 layers each with 9 unit cubes.

FIGURE 12.

Hex found patterns in the plane by counting vertices and unit segments and unit squares in a subdivided square. In space, we can count vertices, unit segments, unit squares, and unit cubes in a subdivided cube (Figure 13).

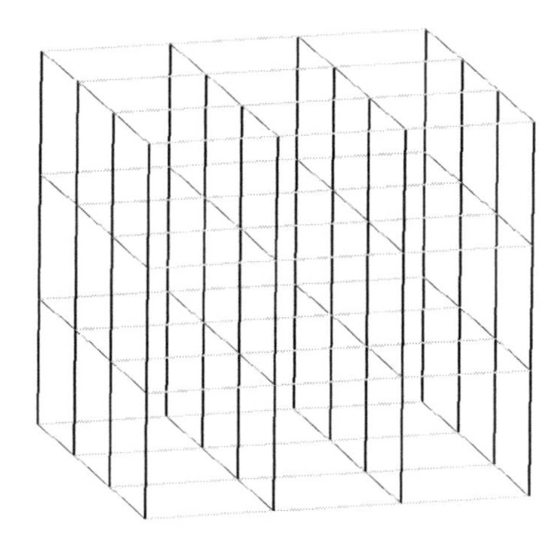

FIGURE 13.

As in the case of the plane, we are aided in our counting of the edges by considering each direction separately. As before, we get some good information by looking at a simpler case (Figure 14). For a $2 \times 2 \times 2$ subdivided cube, we have 8 unit cubes.

Next, we have 27 vertices, 9 in each of 3 horizontal planes.

We have 54 unit segments, 12 in each of the three horizontal planes and 18 vertical unit segments, 2 for each of the 9 vertices in the horizontal plane.

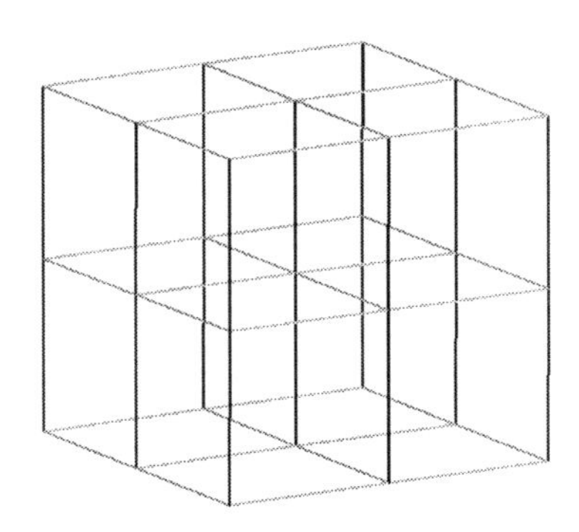

FIGURE 14.

Hardest to count are the unit squares. There are 3 horizontal planes, each with 4 squares, and 6 vertical planes, each with 4 squares. We have a total of 36 unit squares.

Once again, we can enter our numbers in a table and see what patterns we can recognize.

Number of Subdivisions	Verticies (V)	Unit Segments (E)	Unit Squares (S)	Unit Cubes (C)
1	8	12	6	1
2	27	54	36	8
3				

Number of Subdivisions	V + E + S + C	V − E + S − C
1	27	1
2	125	1
3		

CONCLUSION

Modern developments in computer graphics, in particular, the use of color, animations, and interactive materials, can change the way we approach the relationship between algebra and geometry, in classical two-dimensional topics as well as in less familiar analogues of these topics in solid geometry. These opportunities for exploration will definitely increase in the future, and become a more and more important part of curricula at all levels.[1]

REFERENCES

Alsina, C., & Roger N. (2006). *Math made visual: Creating images for understanding mathematics.* Washington D.C.: Mathematical Association of America .

Banchoff, T. (2008). *Algebraic thinking and geometric thinking.* In Greenes, C. E. (Editor). *Algebra and algebraic thinking in school math: NCTM seventieth yearbook* (pp. 99–112). Reston VA: NCTM.

Battisa, M. T., & Clements, D. H. (1998). Students' spatial structuring and enumeration of 2-D arrays of squares. *Journal for Research in Mathematics Education, 29*(5), 503–532.

Caplan, S. (Producer), Johnson, D., & Travis, J. (Directors). (2007). *Flatland: The movie* [Motion picture]. United States: Flatworld Productions, LLC..

[1] Illustrations and animations for this article are due to Michael Schwarz, a 2008 graduate from Brown with a Math-Physics concentration, using software developed by undergraduate students under the direction of the author, with support from the National Science Foundation and from Brown University.

CHAPTER 13

THOUGHTS ON ELEMENTARY STUDENTS' REASONING ABOUT 3-D ARRAYS OF CUBES AND POLYHEDRA

Michael T. Battista

A BRIEF SUMMARY OF MY RESEARCH ON 3-D GEOMETRY

Context

I started my research on 3-D geometry in preparation for writing several 3-D geometry curriculum units for the *Investigations in Number, Data, and Space* elementary mathematics curriculum. In this initial research (Battista & Clements, 1996), we conducted interviews and teaching experiments with individual and small groups of students in grades 3, 4, and 5. My research continued as these instructional units were field-tested by classroom teachers (Battista, 1999; Battista & Clements, 1998). For instance, in one study (Battista, 1999), I studied pairs of students participating in a fifth-grade class that was field-testing the *Investigations* fifth-grade unit on 3-D geometry (Battista & Berle-Carman, 1996). I have revised and consolidated

Future Curricular Trends in School Algebra and Geometry: Proceedings of a Conference. pages 183–199

many of these research ideas in my development of and research for the *Cognition Based Assessment* system for elementary school mathematics (Battista, 2004). So my research on 3-D geometry has been fully integrated with my curriculum development efforts.

In this paper, I briefly describe highlights of my past research, and I describe some new data on how students construct spatial meaning for polyhedra and some new directions for investigating the mental processes for students' construction of meaning for 3-D shapes.

PAST RESEARCH

Most of my research has focused on students' reasoning about volume measurement. In particular, we investigated how students solved problems such as the following (problems were presented both pictorially, and concretely with interlocking cubes, with no disassembly permitted):

Task. How many unit cubes will it take to make the building below? The building is completely filled with cubes, with no gaps inside.

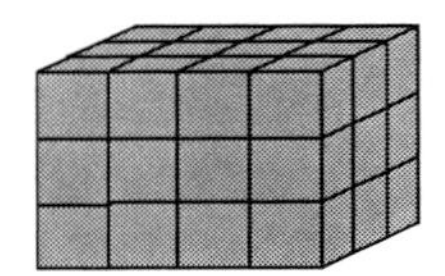

INITIAL FINDINGS ON STUDENTS' ENUMERATION OF 3-D ARRAYS OF CUBES

In a study of third- and fifth-grade students (Battista & Clements, 1996), we classified the strategies that students employed during individual interviews into five broad categories.

Category A: The student conceptualized rectangular arrays of cubes as layers. Cubes could be enumerated by counting (individually or by skip counting), adding, or multiplying.

Category B: The student conceptualized the set of cubes as space-filling, attempting to count all cubes in the interior and exterior, but did not consistently organize the cubes into layers. Strategies ranged from random and unorganized counting as if the cubes were unorganized as depicted in Fig X, to inconsistent but accurate grouping procedures (see Fig Y).

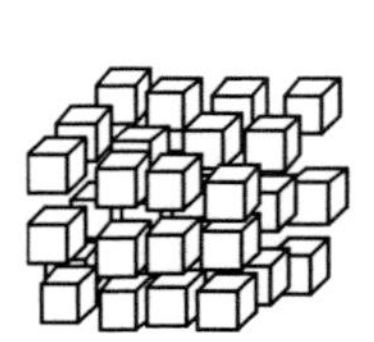

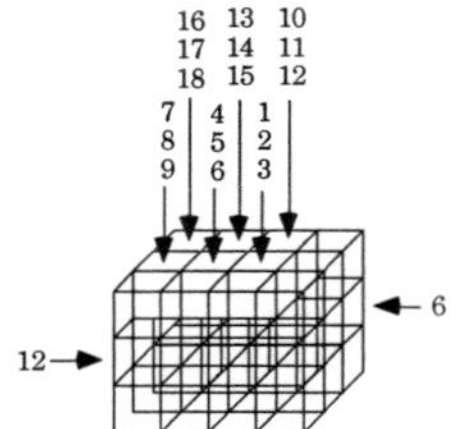

Category C: The student conceptualized the set of cubes in terms of its faces; he or she counted all or a subset of the visible faces of cubes (as seen either from the picture viewing perspective or from all perspectives). For example, for the cube array on the preceding page, Jeff counted all cube faces that appear on the six sides, double-counting edge cubes and triple-counting corner cubes: 12 + 12 + 9 + 9 + 12 + 12.

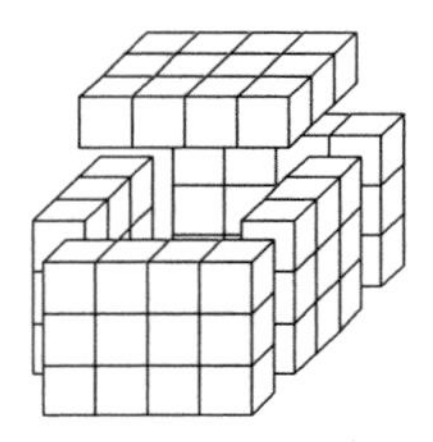

Category D: The student explicitly used the formula L × W × H, but with no indication that he or she understood the formula in terms of layers.

Category E: Other. This category includes strategies such as multiplying the number of squares on one face by the number on another face.

Summary data. Of the students who we interviewed, about 60% of third graders and about 20% of fifth graders used a strategy that suggests that the student saw the arrays as consisting of their outer faces. Only 7% percent of 3rd graders and 29% of fifth graders correctly used a layering strategy for all three problems. No student used the volume formula meaningfully. Also, 64% of the third graders and 21% of the fifth graders double-counted cubes at least once on the three interview questions. These percents are likely to be underestimates for the general population because these third and fifth grade students were above average in mathematics (average = 80th percentile). Indeed, Ben-Chaim et al. (1985) found 39% of students in grades 5-8 counting cube faces, and 16% counting cubes on faces, both of which would be classified roughly as Category C. *So significant numbers of students in the elementary and middle school seem to focus on the faces of the rectangular arrays.* In the next section, I briefly describe research that investigates more deeply students' construction of meaning of 3-D arrays of cubes (Battista, 1999, 2004; Battista & Clements, 1996).

UNDERLYING MENTAL PROCESSES (Battista, 1999, 2004)

In addition to the process of abstraction, four additional cognitive processes are essential for meaningful enumeration of arrays of cubes: forming and using mental models, spatial structuring, units-locating, and organizing-by-composites.

In the *forming and using mental models process,* individuals create and use imagistic or recall-of-experience-like mental representations that have structures isomorphic to the perceived structures of the situations they represent (Battista, 1999). Mental models consist of integrated sets of abstractions that are activated to visualize, comprehend, and reason about situations that one is dealing with in action or thought.

In the *spatial structuring process,* individuals abstract an object's composition and form by identifying, interrelating, and organizing its components. For instance, to spatially structure a rectangular array of cubes, one might see it as a set of columns, or a set of layers. The *units-locating process* locates cubes by coordinating their locations along the dimensions that frame an array. Students can meaningfully enumerate arrays of cubes only if, through the process of abstraction, they have developed properly structured mental models that enable them to correctly locate and organize the cubes.

The *organizing-by-composites process* combines an array's basic spatial units (cubes) into more complicated composite units that can be repeated or iterated to generate the whole array. For instance, in a 3-D array, the cubes in a horizontal layer can be grouped into a layer to form a spatial composite that can be iterated vertically to generate the array.

LEVELS OF SOPHISTICATION IN STUDENTS' ENUMERATION OF 3-D ARRAYS (Battista, 2004)

The four cognitive processes described above will now be used to describe levels of sophistication in students' understanding of volume measurement. Additional processes needed to construct mental models of 3-D arrays of cubes, in particular the oft-mentioned *coordination* process, will be discussed in the section following this one.

Level 1: Absence of Units-locating and Organizing-by-composites Processes

Students do not organize units into spatial composites, and, because they do not properly coordinate spatial information, their mental models of cube arrays are insufficient to locate all the units in arrays.

> *Task: The student is asked how many cubes are needed to completely fill the box shown below.*

Bob counted the 8 cubes shown in the box, then pointed to and counted 6 imagined cubes on the box's left side, 4 on the back, 4 on the bottom, and 5 on the top. His units-locating process was insufficient to create an accurate mental model of the cube array.

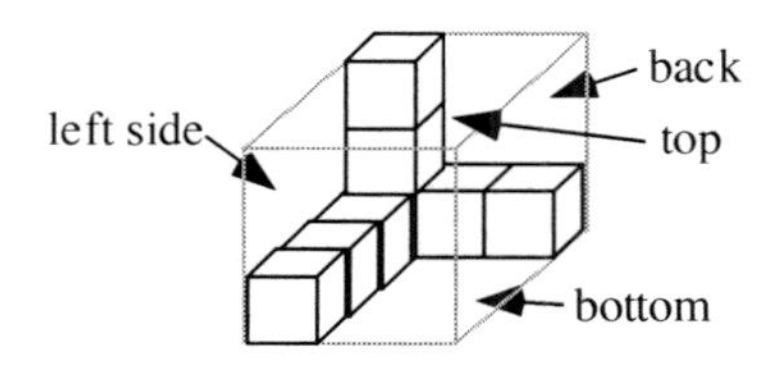

Level 2: Beginning Use of the Units-locating and the Organizing-by-composites Processes

Students not only start to spatially structure arrays in terms of composite units, their emerging development of the units-locating process produces mental models sufficient for them to recognize equivalent composites. For example, for the building shown below—Fred counted 12 cubes on the front, then immediately said there must be 12 on the back; he counted 16 on the top, and immediately said there must be 16 on the bottom; finally, he counted 12 cubes on the right side, then immediately said there must be 12 on the left side.

In each case, after counting the cubes visible on one side of the building, Fred *inferred* the number of cubes on the opposite side, clear evidence that he was organizing cubes into composites and that he was using the units-locating process to relate these composites spatially and numerically.

Also, note the ubiquitous "double-counting" error that Fred made on this task, providing clear evidence that insufficient coordination of views causes major difficulty in the units-locating and units-identification processes. Because he could not properly coordinate what he saw on the different sides of the building, Fred failed to realize when adjacent cube faces were part of the same cube.

Level 3: Units-locating Process Becomes Sufficiently Coordinated to Recognize and Eliminate Double-Counting Errors

A major breakthrough in thinking occurs when a student's units-locating process coordinates single-dimension views (e.g., top, side, front) and integrates them into a mental model that is sufficient to recognize the same unit from different views. This refined mental model enables students to eliminate double-counting errors caused by insufficient coordination.

> *Task: The student is asked to predict how many cubes it takes to completely fill the box shown below.*

As shown in the figure, Juan coordinated spatial information sufficiently to avoid double-counting edge cubes. However, his coordination was still insufficient to build a mental model that properly located interior cubes.

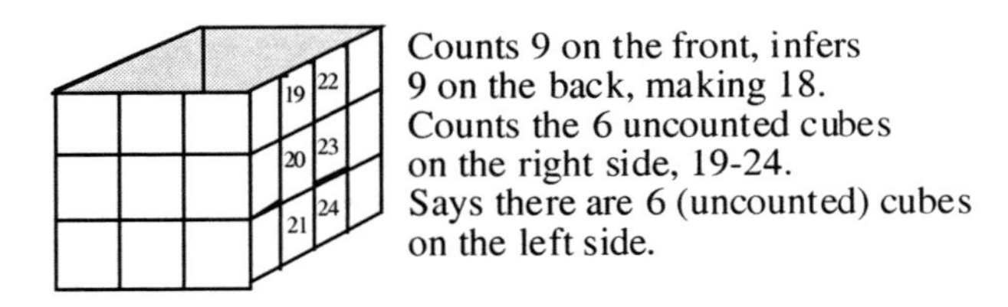

Level 4: Use of Organizing-by-composites Process to Structure an Array as Layers, But Insufficient Coordination for Iteration

Students structure arrays in terms of layers, producing more powerful and efficient mental models of arrays.

> *Task: The student is shown a picture of a 5 by 3 by 4 cube array and is asked to build the bottom layer for the array, then to predict how many cubes it takes to make the whole array. Randa built the 5 by 3 bottom layer and said that it contained 15 cubes.*

When asked how many cubes were in the entire building, she counted from the bottom up on the picture, but continued to count on the top, getting 7 layers (see below). She gave an answer of 105 cubes. Randa could not coordinate the horizontal layers with the third dimension that was the prism's height.

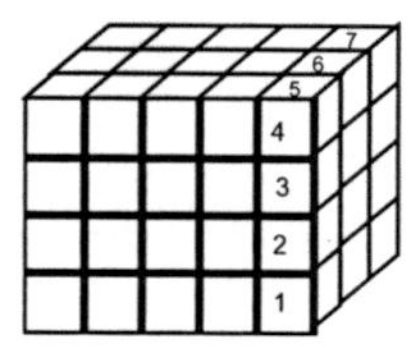

Level 5: Use of Units-locating Process Sufficient to Correctly Locate All Units, But Layers Are Not Used

This is the first level in which the units-locating process is sufficient to create a mental model that correctly locates all cubes in an array. However, although students sometimes get correct answers, because they inefficiently or inconsistently organize arrays into composites, they quite frequently lose their place in counting/adding and make enumeration errors. Furthermore, students' structuring and enumeration strategies are not generalizable and are inadequate for large arrays.

> *Task: The student is asked to predict the number of cubes it takes to make a 4 by 3 by 3 cube building shown in a picture (the student is told that the building is completely filled with cubes inside).*

As shown below, Mary counted the cubes visible on the front face (12), then counted those on the right side that had not already been counted (6). She then pointed to the remaining cubes on the top, and for each, counted cubes in columns of three: 1, 2, 3; 4, 5, 6; ... 16, 17, 18. She then added 18, 12, and 6.

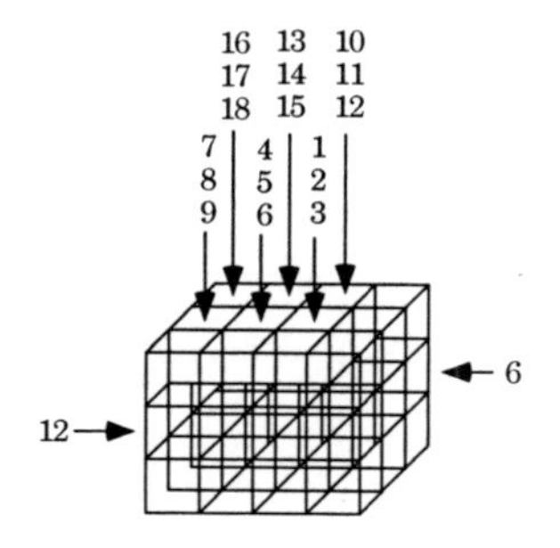

Level 6: Complete Development and Coordination of Both the Units-locating and the Organizing-by-composites Processes

Students' mental models fully incorporate a layer structuring so that students can accurately reflect on and enumerate 3-D array of cubes.

Level 7. Students' Spatial Structuring and Enumeration Schemes become Sufficiently Abstract so that Students Can (a) Understand the Connection between Numerical Procedures and Spatial Structurings, and (b) Generalize their Reasoning to non-Cubic Rectangular "Packages"

ESSENTIAL MENTAL PROCESSES: COORDINATION, INTEGRATION, AND DEVELOPING PROPER MENTAL MODELS (from Battista & Clements, 1996)

According to Morss (1987), children initially perceive 3-D configurations without reference to specific vantage points or perspectives. That is, they first conceive of the configurations in terms of a "medley of viewpoints," as an uncoordinated set of images. Later in the course of development, the child *constructs* the notion of single viewpoints per se, recognizing them as such. Even later in development, after the construction of single perspectives, the child becomes capable of coordinating perspectives. This developmental trend is evidenced in children's drawings, in which young children draw 3-D objects as a single face or an uncoordinated set of faces, but older children employ a single, coherent perspective, with visual realism being achieved when students finally learn how to pictorially represent their coordination of the different included views (Mitchelmore, 1980).

A MEDLEY OF VIEWS

Applying Morss' developmental theory of perspective thinking to students' conceptualization of 3-D cube arrays suggests that students' initial conception of 3-D rectangular arrays is an uncoordinated set, or medley, of views (especially orthogonal views).

COORDINATION

To move beyond the medley-of-views conception of cube arrays, students must *coordinate* orthogonal views. That is, they must recognize how these views are spatially interrelated. Coordinating views of a 3-D array requires that cube faces depicting the same cubes be recognized as such (which is not the case when students double count cubes along the edges of a rectangular array).

INTEGRATION

But students must do more than coordinate the views; they must integrate them. To *integrate* views of a 3-D object, students must coordinate the views, then construct a single, coherent mental model of the object that possesses these views. The above data suggests, however, that many students are unable to enumerate the cubes in a 3-D array because they cannot coordinate the separate views of the array and integrate them to construct one coherent mental model of the entire array, inside and out.

As an example, consider fifth-grader RA's performance on the following task:

> RA was shown a picture of a box with the length, width, and height labeled and was told, "This box contains 3 cubes along the bottom, 3 up from here to here, and 4 from here to here [pointing appropriately at the box picture]." She was then asked, "How many cubes does it take to completely fill the box? Draw what the cubes look like on the outside of the box."

After about ten to fifteen minutes and many erasures, RA's drawing looked like the figure below, showing a clear lack of coordination of spatial information. (For instance, the right side shows 5 horizontal rows; the front shows only 3.)

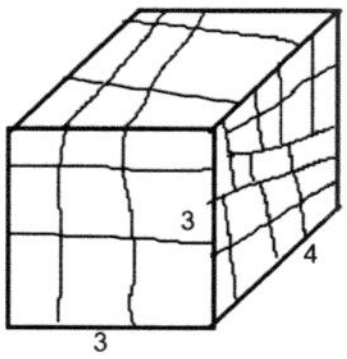

3
4
3

To corroborate RA's lack of coordination, RA was given two orthogonal views of a simple five-cube object shown below and asked to build the object. She built one cube configuration that looked like the figure on the left, and another that looked like the figure on the right. She said that the task was impossible until the interviewer showed her the correct configuration, which she immediately recognized as correct.

> *Int: I've put some cubes together to make an object that looks like this [pointing to the left figure] if you look at it straight on—like you could see it through my hand [illustrating with hand]—and like this [pointing to the figure on the right] if you turn it [turning hand 90°].*

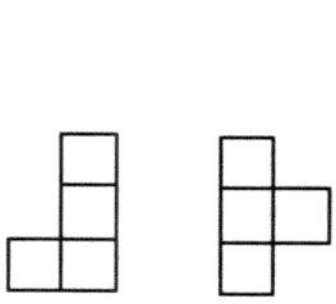

In a subsequent administration of this task to a group of fifth graders, 18 of 47 students built each of the two views shown and simply affixed them together. Nine of these students gave this combination as the solution. They neither coordinated nor integrated the views. The other nine knew that there was something wrong with their answer but could not figure out what. They seemed to be able to coordinate the views but not to integrate them into a coherent whole.

EXTENDING THE ANALYSIS: NEW DATA ON MEDLEY OF VIEWS

I now describe an example that illustrates that the medley-of-views conceptualization also occurs in somewhat different contexts.

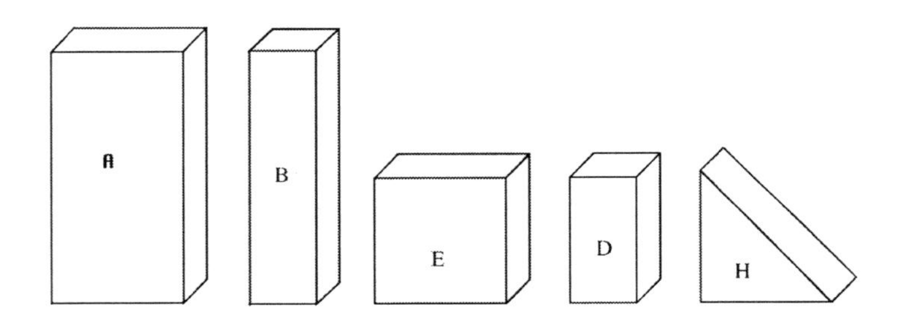

Wooden Blocks

Int: How does the amount of wood in one of these [block E] compare to the amount of wood in one of these [block B]?
CH: This one [B] is just as fat as this one [E], but this one [E] is shorter, and this one [B] is longer.
Int: How many of these [B] does it take to make one of these [A]?
CH: I think 2 [checks with blocks], yeah, 2.
Int: How many of these [E] does it take to make one of these [A]?
CH: 2.
Int: So, how do these two compare, this one [B] and this one [E]?
CH: Oh, they're the same because if you can fit 2 of these [E] on one of these [A], and 2 of these [B] on one of these [A], then they're the same.

In this case, CH, with some guidance from the interviewer, used the transitive property to conclude that the amount of wood in two blocks was the same, even though the blocks had different shapes. (The amount of wood serves as a proxy for volume.)

After doing several similar tasks, the interviewer asked CH how many of block D it takes to make block A. CH makes face-to-face comparisons and matching and concludes that it takes 14. (See below.)

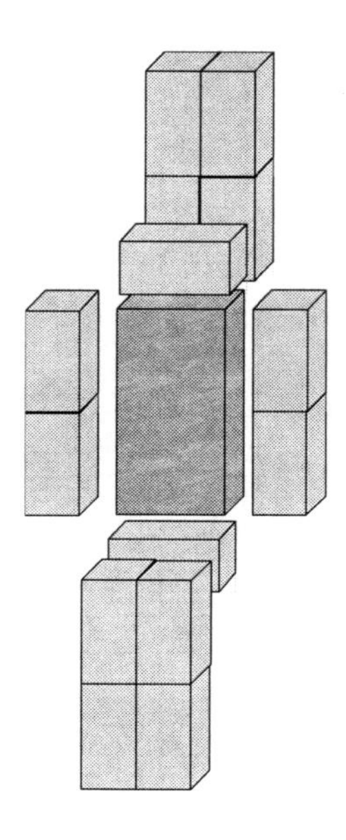

The interviewer has CH build A from Ds and CH sees that it takes 4. After building, CH also sees that it takes 4 Hs to make A. Asked to compare the amount of wood in H and D, CH concludes that H is bigger. He compares the 3 rectangle faces of H with the rectangle faces of D and says that 2 are the same but one is longer, so H has more wood. (See below.)

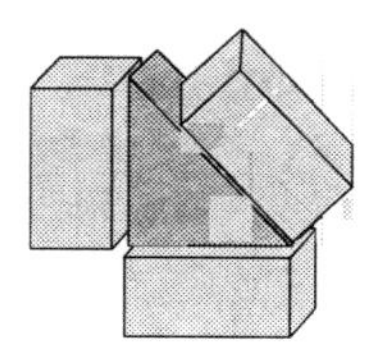

For the present discussion, what is interesting is that, as CH moves from holistic visual comparison to parts-based comparison, he seems to lose sight of the whole polyhedron as a structured integration of its parts, falling into the uncoordinated medley-of-views conception. He makes errors in thinking about the amount of wood (volume) in the blocks, and in actually thinking about how to build one kind of block from others, because he focuses on faces in a way that does not preserve the 3-D structure of the blocks. As before, I conjecture that this medley of views conception is a result of insufficient coordination and integration processes.

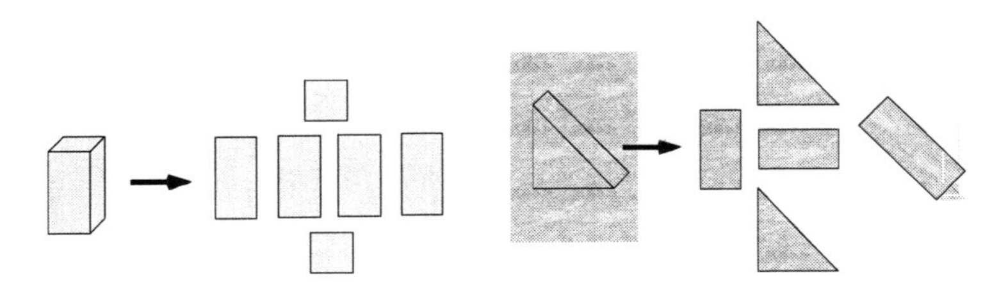

DELVING DEEPER INTO COORDINATION AND INTEGRATION: RELATED RESEARCH

Given students' difficulties with coordinating and integrating views of 3-D objects, it is essential to better understand the coordination and integration processes. One way to do this is to examine other situations in which spatial coordination and integration are required. One such situation is when information is presented about a phenomenon using several diagrams. For instance, consider map books in which detailed maps of sections of a city appear on different pages in a book, or even consider the use of GPS systems in which only a portion of a map is shown at a time, giving rise to the "keyhole" effect (Kim & Hahn, 1997). Although I have yet to find research on these situations that sheds light on the coordination and integration processes, I have found some relevant research on another situation which has some relevance.

INTEGRATION IN UNDERSTANDING A SET OF RELATED DIAGRAMS

Kim and Hahn (1997) and Kim, Hahn, and Hahn (2000) have studied how college engineering students make sense of multiple diagrams about business systems. They postulate two related processes. To perform the *perceptual integration process,* individuals "need to locate the diagram to look at next and need to recognize the relevant items in the target diagram" (Kim & Hahn, 1997, p. 377). The *conceptual integration process* "consists of generating and refining hypotheses by linking the information inferred from different diagrams" (p. 377). These researchers further hypothesize that to facilitate the perceptual integration process, diagrams must provide visual cues that indicate how an item in one diagram is related to items in other diagrams. To facilitate the conceptual integration process, they hypothesize that the set of diagrams must provide a broader context that indicates how the various diagrams fit into the overall picture (e.g., a map of the whole city with rectangles superimposed on it that indicate where various page-maps occur).

In the case of 3-D arrays of cubes, Kim's perceptual integration seems similar to what I have called coordination—students must recognize which cube faces are part of the same cube. And Kim's conceptual organization seems similar to my reference to the construction of mental models that integrate views orthogonal to array faces.

One might think that giving students a cube building or isometric view diagram provides the overall picture that should relate the orthogonal views. However, our data indicates that this is insufficient for many students, so it is not the diagram per se, but the construction of an integrated mental model that is critical for integration. And importantly, the work of Kim et al. suggests that generating and testing hypotheses is critical for mental model construction.

Interestingly, the introductory activity in the fifth grade *Investigations* curriculum unit *Containers and Cubes* (Battista & Berle-Carman, 1996) attempted to provide support for both coordination and integration in ways consistent with the Kim et al. research. In this activity, students predict the number of cubes needed to fill open boxes shown in pictures that showed cube faces, both as boxes and nets for the boxes (see the example below). Students test their predictions by building the boxes and filling them with cubes.

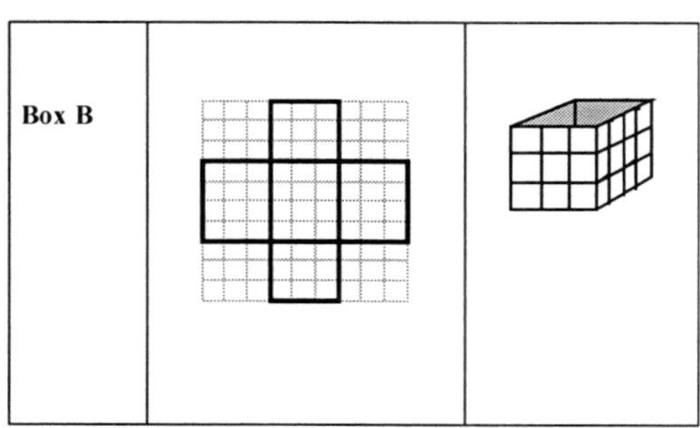

Making a box and filling it with cubes encourages coordination because it encourages students to see how the squares on the sides of the boxes are related to cube faces and how the box sides are related to the faces of the completed box/cube array. The predict-and-check approach used in this unit also specifically requires students to generate and test hypotheses, consistent with the Kim research cited above. However, for students who do not coordinate on their own, the Kim research suggests that a good prompt for edge cubes might be, "Show me on this box side where this cube (from another side) is."

SUPPORTING THE INTEGRATION PROCESS

I now describe two methods that might help students develop more integrated models of 3-D shapes—using color and using computer animations. Research should investigate both methods.

USING COLOR

The Kim research indirectly suggests several possibilities for supporting students' integration processes that would be interesting to test empirically. For instance, using color (which can only be hinted at in the grayscale illustrations below) might promote the perceptual integration that supports conceptual integration. In the figure below, the colors can provide perceptual clues that help students coordinate information contained in the three orthogonal views about the location of edge and corner cubes. For instance, a color version of the building below on the left might make much more salient to students the fact that the upper right square on the front face is part of the upper right corner cube.

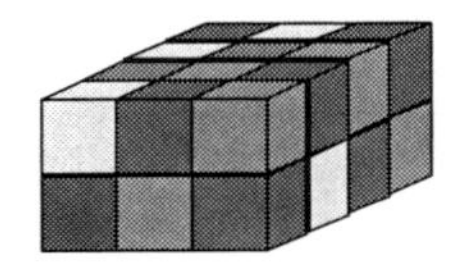
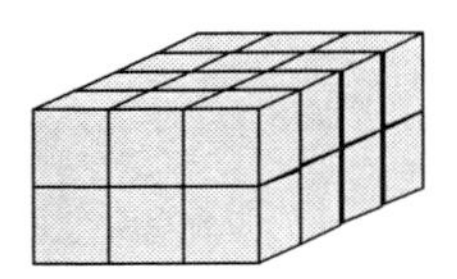

Color might also help students with a task that explicitly deals with coordination and integration of views. I hypothesize that doing such tasks helps students develop proficiency with mental processes needed to construct appropriate mental models of 3-D cube arrays. As shown on the next page, coloring schemes might provide extra clues needed for perceptual integration of the three orthogonal projections to create an appropriate mental model of the 3-D cube configuration. It would be worthwhile for research to investigate this hypothesis.[1]

[1]Note for the figure on the left, (a) the front, top, and right-side views of the configuration are all supposed to be different colors, with each face shown in one of the views the same color; and (b) this coloring scheme works for a picture, but not for plastic cubes because instructional cubes generally have all six of their faces the same color.

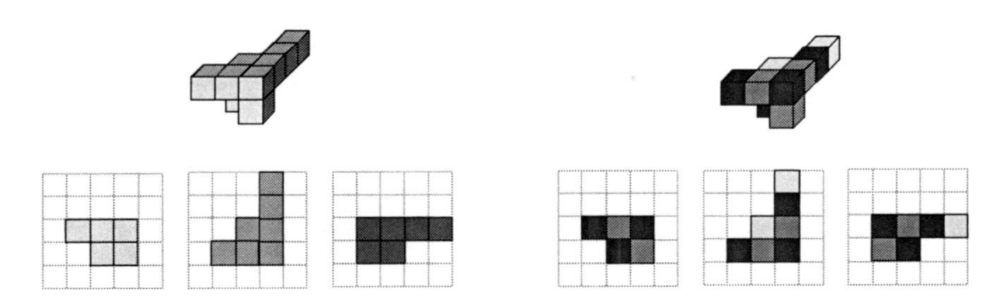

USING COMPUTER ANIMATIONS

Another possible integration-supporting activity is the use of motion in computer animations. I hypothesize that motion can help students establish coordinating connections between corresponding cubes in orthogonal views as these views are continuously changed (continuity seems to support maintenance of the coordinating links between components of views). For instance, the *Space* module in *Middle School Maths* website (http://members.westnet.com.au/molinasantos/space.htm) provides both motion and explicit attention to orthogonal views of cube configurations, encouraging hypothesis creation and testing in the construction of mental models for cube configurations.

As an example, consider the situation presented below. To construct the configuration depicted in the top, front, and side views, you can add and subtract cubes on the gray computer screen grid, and you can rotate your cube configuration on the screen to inspect the top, front, and side views.

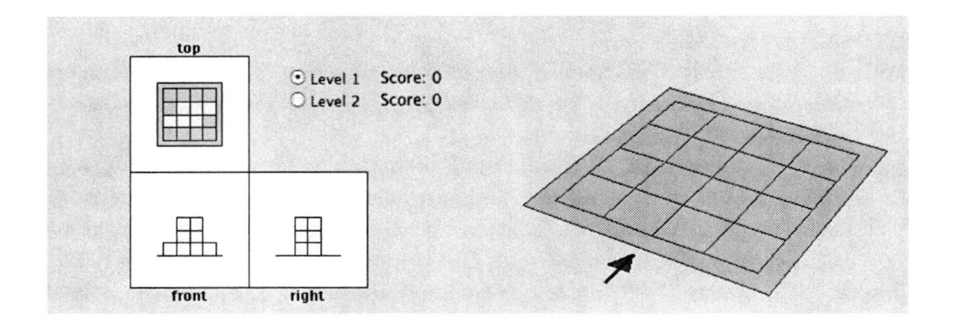

To illustrate, students might begin their solution attempt by adding the cubes shown below (to make the top view).

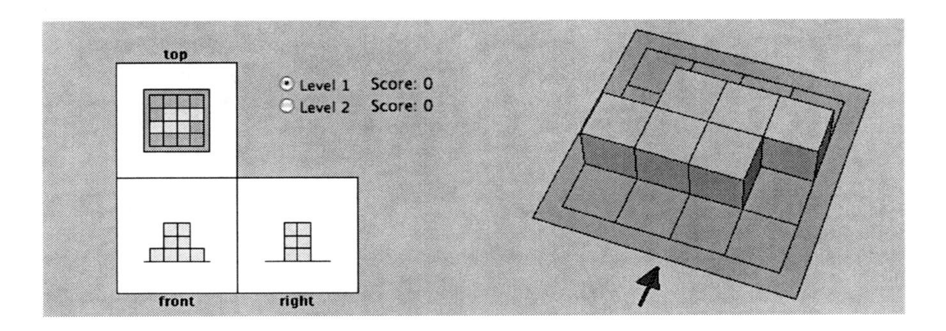

As they rotate this cube configuration on the grid, it immediately "pops out" as a 3-D, not 2D, figure. And by rotating the configuration, they can see that, although its top view is correct, its front and side views are not yet correct. They can they generate and test a new hypothesis about the required configuration. Presumably, rotation of the configuration on the computer screen can help students coordinate views and maintain an integrated mental model of the configuration. It would be interesting to investigate whether using such software helps students to better use strictly mental visualization of such cube configurations.

REFERENCES

Battista, M. T. & Berle-Carman, M. (1996). *Containers and cubes.* Palo Alto, CA: Dale Seymour Publications.

Battista, M. T. & Clements, D. H. (1996). Students' understanding of three-dimensional rectangular arrays of cubes. *Journal for Research in Mathematics Education,* 27(3), 258–292.

Battista, M. T. (1999). Fifth graders' enumeration of cubes in 3-D arrays: Conceptual progress in an inquiry-based classroom. *Journal for Research in Mathematics Education, 30*(4), 417–448.

Battista, M. T., & Clements, D. H. (1998). Students' understanding of 3-D cube arrays: Findings from a research and curriculum development project. In R. Lehrer & D. Chazan (Eds.), *Designing learning environments for developing understanding of geometry and space* (pp. 227–248). Mahwah, NJ: Lawrence Erlbaum.

Battista, M.T. (2004). Applying cognition-based assessment to elementary school students' development of understanding of area and volume measurement. *Mathematical Thinking and Learning, 6*(2), 185–204.

Ben-Chaim, D., Lappan, G., & Houang, R. T. (1985). Visualizing rectangular solids made of small cubes: Analyzing and effecting students' performance. *Educational Studies in Mathematics, 16,* 389–409.

Kim, J. & Hahn, J. (1997). Reasoning with multiple diagrams: Focusing on the cognitive integration process. *Proceedings of the 19th Annual Conference of the Cognitive Science Society,* (pp. 376–381). Mahwah, NJ: Lawrence Erlbaum.

Kim, J., Hahn, J., & Hahn, H. (2000). How do we understand a system with (so) many diagrams? Cognitive integration processes in diagrammatic reasoning. *Information Systems Research, 11*(3), 284–303.

Mitchelmore, M. C. (1980). Prediction of developmental stages in the representation of regular space figures. *Journal for Research in Mathematics Education, 11,* 83–93.

Morss, J. R. (1987). The construction of perspectives: Piaget's alternative to spatial egocentrism. *International Journal of Behavioral Development, 10*(3), 263–279.

PART IV

LINKING ALGEBRA AND GEOMETRY

CHAPTER 14

LINKING GEOMETRY AND ALGEBRA IN THE SCHOOL MATHEMATICS CURRICULUM

Keith Jones

INTRODUCTION

This paper could have been entitled *Linking Algebra and Geometry in the School Mathematics Curriculum.* After all, algebra does come before geometry in the dictionary. Yet there are a number of reasons why it might be advantageous to begin this paper with these two components of mathematics in reverse alphabetic order: it switches attention to geometry (rather than bolstering the tendency for algebra to dominate the school mathematics curriculum); it implies that geometry can provide insight into other aspects of mathematics; and it is indicative of how the development of digital technologies has seen a resurgence in interest in geometry and in techniques for visualizing mathematics (Jones, 2000, 2002). For these reasons, and more, the focus of this paper is linking geometry and algebra—and how, through such linking, the mathematics curriculum (and hence the teaching and learning experience) might be strengthened.

Future Curricular Trends in School Algebra and Geometry: Proceedings of a Conference. pages 203–215

THE RELATIONSHIP BETWEEN GEOMETRY AND ALGEBRA

That algebra can tend to dominate the school mathematics curriculum is apparent in many ways, one example being the work of the U.S. National Mathematics Advisory Panel which was directed to focus on "the preparation of students for entry into, and success in, algebra" (U.S. National Mathematics Advisory Panel, 2008, p. 8). Yet it is worth reflecting on the words of people like Coxeter, Bell, and Atiyah (an ABC of renowned mathematicians, taken in reverse alphabetic order). It was Coxeter, the famous geometer, who replied with the following advice when asked what would most improve upper secondary or college level mathematics teaching: "I think that, by being careful, we could probably do the same amount of calculus and linear algebra in less time, and have some time left over for nice geometry" (Coxeter quoted in Logothetti & Coxeter, 1980, p. 16).

The mathematician Eric Bell noted that "With a literature much vaster than those of algebra and arithmetic combined, and at least as extensive as that of analysis, geometry is a richer treasure house of more interesting and half-forgotten things, which a hurried generation has no leisure to enjoy, than any other division of mathematics" (Bell quoted in Coxeter & Greitzer, 1967, p. 1).

At his Fields Lecture at the World Mathematical Year 2000 Symposium (Toronto, Canada, June 7–9, 2000), the celebrated mathematician Michael Atiyah argued that "...spatial intuition or spatial perception is an enormously powerful tool and that is why geometry is actually such a powerful part of mathematics—not only for things that are obviously geometrical, but even for things that are not. We try to put them into geometrical form because that enables us to use our intuition. Our intuition is our most powerful tool" (Atiyah, 2001).

In the history of mathematics there has, it seems, been a somewhat (and sometimes) uneasy relationship between geometry and algebra (Atiyah, 2001; Charbonneau, 1996; Giaquinto, 2007; Kvasz, 2005). According to Atiyah (2001), fundamental to what can seem like a dichotomy is that "algebra is concerned with manipulation in time, and geometry is concerned with space. These are two orthogonal aspects of the world, and they represent two different points of view in mathematics. Thus the argument or dialogue between mathematicians in the past about the relative importance of geometry and algebra represents something very, very fundamental". Yet while algebra provides powerful techniques for mathematics, Atiyah sees a danger that "when you pass over into algebraic calculation, essentially you stop thinking; you stop thinking geometrically, you stop thinking about the meaning".

These are some of the reasons for focusing on linking geometry and algebra, for recognising the important role that geometrical thinking has in mathematics, and for strengthening the teaching and learning of mathematics through finding ways of building on students' spatial intuition and spatial perception. That it remains vital to do these things might be surmised from considering the case of the school mathematics curriculum in England.

THE SCHOOL MATHEMATICS CURRICULUM: THE CASE OF ENGLAND

The introduction, in 1988, of a statutory national curriculum in England cemented existing UK practice that "while it is convenient to break mathematics down ...[into areas such as number, algebra, geometry, statistics].., it is important to remember that they do not stand in isolation from each other (UK DES, 1988, p. 3). In this way, mathematics in UK schools is generally presented as an integrated subject, although students may well experience a curriculum diet of mathematics taught as a series of separate topics (of algebra, geometry, and so on) of varying lengths (of perhaps four to six weeks each). Parallel to the introduction of a statutory national curriculum, a system of national testing for students aged 7, 11, and 14 was instigated, augmenting existing national testing at 16 and 18.

In the period from 1995–2000, this form of statutory curriculum and national testing became more and more entrenched, leading to an increase in forms of school accountability through the publication of, for example, "league tables" of schools (based on their national test results). At this time, also, international comparisons such as TIMSS began to have an increasing impact; so much so that the UK Government launched its *National Numeracy Strategy* in 1998 (Department for Education and Employment, 1998a; 1998b; 1999). This numeracy strategy sought to address perceived weaknesses in the teaching of mathematics, particularly at the elementary school level, and focused primarily on skills of calculation and computation. Geometry received hardly a mention (Jones & Mooney, 2003). At the same time, there were emerging concerns about mathematics teaching at the secondary school level, particularly regarding perceived inadequacies in the preparation for proof at University level (London Mathematical Society, 1995; Engineering Council, 1999).

During this period, the International Commission on Mathematical Instruction (ICMI) study on the teaching and learning of geometry was taking place (Mammana and Villani, 1998), with, amongst many other issues, a consideration of the relationship between deductive and intuitive approaches to solving geometrical problems (Jones, 1998) and the nature and role of proof in the context of dynamic geometry software (Hoyles & Jones, 1998).

In the period 2000–2005, the statutory curriculum for England was revised. The revision of the mathematics curriculum included more explicit stipulations regarding proof, and some further encouragement for links within mathematics and across subjects. National testing continued in much the same form, with school accountability in the form of league tables of schools published in the national media becoming even more ingrained. The national *numeracy* strategy was extended into secondary schools as the national *mathematics* strategy (Department for Education and Employment, 2001).

During this period, the UK Royal Society and Joint Mathematical Council instigated a working group on the teaching and learning of geometry from age 11 to 19 (Royal Society, 2001). The report of this working group stressed the far-reaching importance of geometry within and beyond the school mathematics curriculum and was widely welcomed. Amongst the themes of the report were an emphasis on conjecturing and proving, on the importance of spatial thinking and visualizing, plus the benefits of linking geometry with other areas of mathematics, and on the powerful role of digital technology. A number of the report's recommendations have already been enacted within the UK school system, with some being illuminated through a UK Government initiative on algebra and geometry (Qualifications and Curriculum Authority, 2004). This initiative sponsored six modest curriculum development projects, with the overall report stressing that "making connections between different mathematical concepts is important for developing understanding [of mathematics]" (*ibid*, p. 25). The report, in summarising the six individual projects, offered two suggestions of ways of linking geometry and algebra, one being to exploit the capacity of dynamic geometry software to provide novel ways of visualizing algebraic relationships, the second being to use different approaches to tackle the same problem. Such ways of linking geometry and algebra are illustrated below.

Since 2005, the UK has revised its statutory curriculum again. This time, while established school "subjects" remain, there is less emphasis on specifying the curriculum in terms of subjects (Qualifications and Curriculum Authority, 2005). Despite this, the system of national testing and the entrenched school accountability remains (with the continuing use of league tables) even though an increasing volume of evidence suggests such a system narrows the curriculum (to the testing regime) and thence stifles innovation in curriculum and limits teachers' professional autonomy (for an overview of the state of mathematics teaching in the UK, see Ofsted, 2008).

Around this time, ICMI study 17 on technology examined, amongst many other things, the design of digital technologies for different geometries (Jones, Mackrell, & Stevenson, 2009), and the European Union funded projects on the teaching of three dimensional geometry (Christou, Jones, Mousoulides, & Pittalis, 2006) and the teaching of calculus with dynamic geometry software (Zachariades, Jones, Giannakoulias, Biza, Diacoumopoulos, & Souyoul, 2007).

All this suggests that while international comparisons of mathematical achievement can lead to a government being committed to implementing strict regimes of statutory curricula and student testing, it can happen that reports from outside bodies and from research can have an influence such that, over time, some account starts to be taken of under-represented aspects of the mathematics curriculum.

MAKING CONNECTIONS BETWEEN DIFFERENT MATHEMATICAL CONCEPTS

As the QCA report on algebra and geometry (Qualifications and Curriculum Authority, 2004) indicates, one way of linking geometry and algebra is to exploit the capacity of dynamic geometry software to provide novel ways of visualizing algebraic relationships. As an illustration, teachers in a Hampshire school (in England) worked on a project in which their students used

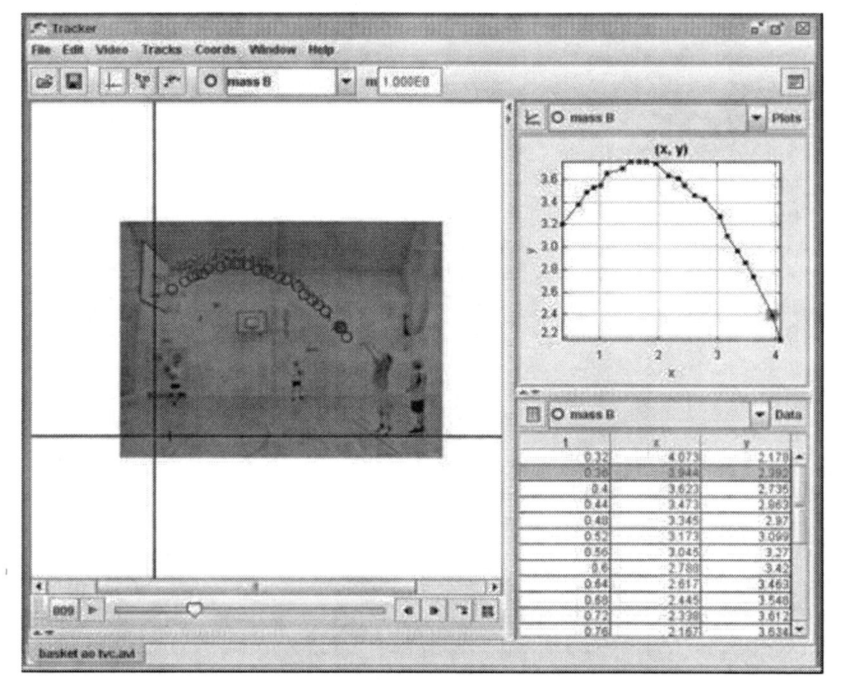

FIGURE 1. Plotting the trajectory of a basketball.

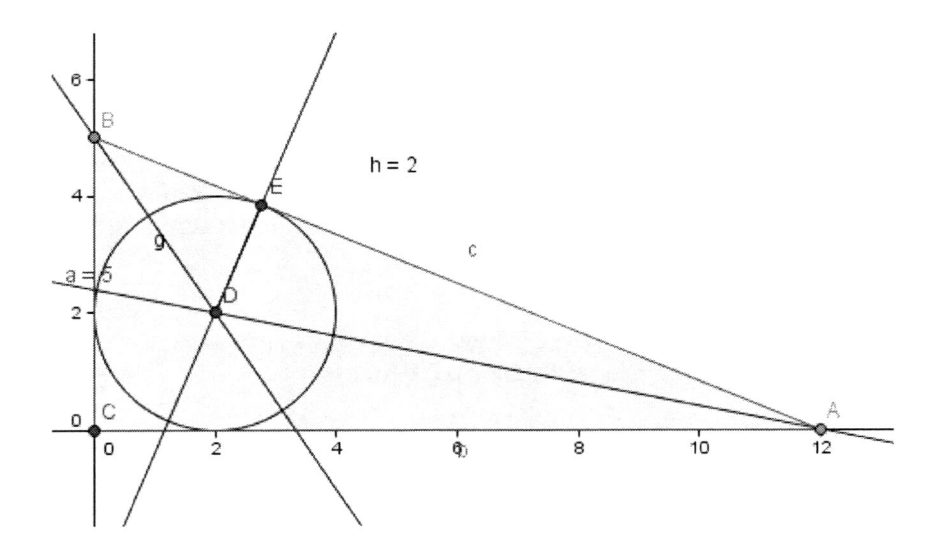

FIGURE 2. Pythagorean triples and integer values of the radius of the incircle.

dynamic geometry software to plot quadratic functions that match the flight of a basketball, providing their students with hands-on experiences of how the various algebraic coefficients affect the shape of the graph.

Another way of linking geometry and algebra is to use different approaches to tackle the same problem. To illustrate this, consider the oft-repeated claim that one of the oldest problems in number theory is to find Pythagorean triples, triples of whole numbers (a, b, c) which fulfill the Pythagorean relation $a^2 + b^2 = c^2$. Yet if a stance is maintained that this is *solely* a problem in number theory, then one outcome is likely to be the omission of the link between Pythagorean triples and the integer size of the radius of the incircle of a right triangle. While not wishing to give too much away to anyone unfamiliar with the construction illustrated in Figure 2, using dynamic geometry to construct the figure and dragging the vertices of the triangle to integer values of the sides of the triangle might suggest a connection to integer values of the radius of the incircle. With, in this way, a conjecture of a theorem being generated, then a small amount of algebra might suffice to prove such a theorem.

It is notable that, in making connections between different mathematical concepts, both of the approaches illustrated in this section of the paper utilize digital technologies.

THE POWER OF GEOMETRY TO BRING CONTEMPORARY MATHEMATICS TO LIFE

A familiar occurrence for many mathematics teachers around the world is students being heard to ask about the usefulness of whatever part of mathematics they are studying. No doubt teachers continually devise inventive attempts to address such questions, yet one thing that might help is to consider how the power of geometry can bring contemporary mathematics to life; examples include double bubbles, black holes, and flags.

A double bubble is a pair of bubbles which intersect and are separated by a membrane bounded by the intersection, as illustrated in Figure 3. It had been conjectured that two partial spheres of the same radius that share a boundary of a flat disk separating two volumes of air use a total surface area that is less than any other arrangement. This equal-volume case was proved

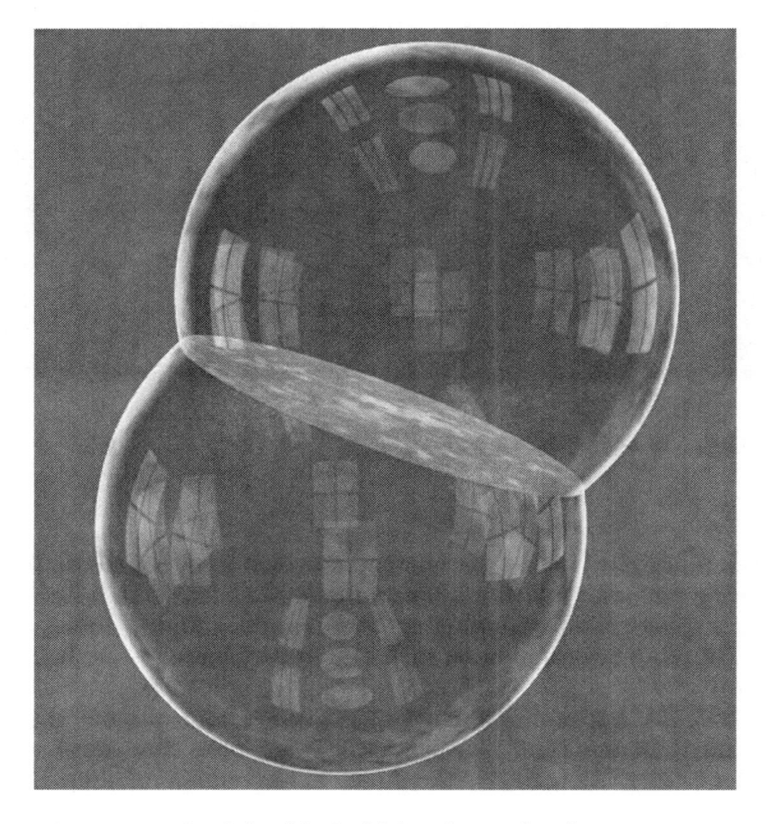

FIGURE 3. A standard double bubble of equal volumes.
© John M Sullivan, jms@uiuc.edu, published by permission.

FIGURE 4. Merging black holes.
© NASA

in 1995. When the bubbles are unequal in size, it has been shown that the separating boundary which minimizes the total surface area is itself a portion of a sphere. Corresponding conjectures about triple bubbles remain open. For more information on such bubble problems, see Brubaker et al (2008).

In 2006, NASA scientists reached a breakthrough in computer modeling that allowed them to simulate what gravitational waves from merging black holes look like (NASA, 2006). The three-dimensional simulations, illustrated by Figure 4, are the largest astrophysical calculations ever performed on a NASA supercomputer.

FIGURE 5. A fluttering US flag.

The design of flags is sometimes mentioned in the school mathematics classroom, perhaps during a topic on symmetry. Yet modeling the movement of a flag mathematically is of interest to mathematicians interested in dynamic systems (such mathematics involves the analytic, asymptotic and numerical solution of non-linear partial singular integro-differential equations with Cauchy Kernels).

In these ways, the power of geometry can be used to bring contemporary mathematics to life. Mentioning these things in the mathematics classroom might mean that learners of mathematics look differently at bubbles, astronomic entities, and flags.

LOOKING TO THE FUTURE

In England, a new curriculum for schools began to be implemented in September 2008. This new curriculum is intended to "give schools greater flexibility to tailor learning to their learners' needs" and as such there is "less prescribed subject content" (QCA, 2007, p. 4). While students are still taught "essential subject knowledge", the new curriculum "balances subject knowledge with the key concepts and processes that underlie the discipline

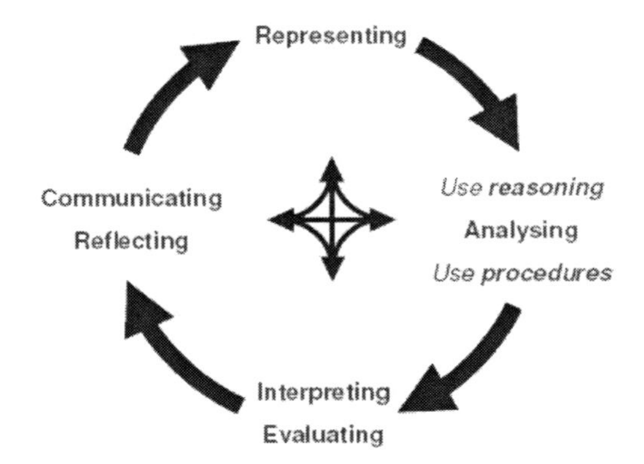

FIGURE 6. Key concepts and processes in the new curriculum for England.

of each subject" (QCA, 2009). In terms of mathematics, these "key concepts and processes" are set out in Figure 6.

This increasing focus on concepts and processes provides new opportunities to ensure that the full potentialities of geometric and algebraic approaches are used for the true benefit of student learning. Already what can appear as the opposing tendencies of geometry and algebra are being blurred in mathematics. For example, Artin, one of the leading algebraists of the 20th century, gave rise to the contemporary use of the term geometric algebra through his book of that title (Artin, 1957). Current applications of geometric algebra include computer vision, biomechanics and robotics, and spaceflight dynamics. Then there is algebraic geometry, the study of geometries that come from algebra. This occupies a central place in contemporary mathematics and has multiple conceptual connections with such diverse fields as complex analysis, topology, and number theory.

The term *concinnity* is most often used for the harmonious or purposeful reinforcement of the various parts of a work of art (with generally the higher the art form, the higher the degree of concinnity). Yet concinnity comes from the Latin *concintas*, meaning skillfully put together, and can apply to any object or situation (even though it is most commonly used in the discussion of music where an example of concinnity might be when the various parts of a piece of music—melody, harmony, rhythm, on so on—reinforce each other).

In the future, we might look for greater concinnity in the mathematics curriculum, especially in terms of the harmonious/purposeful reinforcement of mathematical thinking through the linking of geometry and algebra. Such an approach might be supported by Giaquinto's (2007) view that, from an epistemological perspective, "the algebraic-geometric contrast, so far from being a dichotomy, represents something more like a spectrum".

CONCLUDING COMMENTS

In conclusion, it is worth further reflecting on the words of Atiyah, "...geometry is actually such a powerful part of mathematics—not only for things that are obviously geometrical, but even for things that are not" (Atiyah, 2001).

He went on: "The educational implications of this are clear. We should aim to cultivate and develop both modes of thought. It is a mistake to overemphasize one at the expense of the other and I suspect that geometry has been suffering in recent years. The exact balance is naturally a subject for detailed debate and must depend on the level and the ability of the students involved. The main point that I have tried to get across is that geometry is not so much a branch of mathematics but a way of thinking that permeates all branches" (Atiyah, 1982).

Usiskin (2004) put it this way "the soul of mathematics may lie in geometry, but algebra is its heart"—and, of course, one needs both a heart and a soul. For after all, as is commonly recognized, without geometry, life is pointless.

REFERENCES

Artin, E. (1957). *Geometric algebra.* New York: Wiley.

Atiyah, M. (1982). What is geometry? *Mathematical Gazette,* 66(437), 179-184 [reprinted in Pritchard, C. (Ed) (2003). *The changing shape of geometry: Celebrating a century of geometry and geometry teaching* (pp. 24–29). Cambridge: Cambridge University Press, .

Atiyah, M. (2001). Mathematics in the 20th Century, *American Mathematical Monthly, 108*(7), 654-666.

Brubaker, N. D., Carter, S., Evans, S. M., Kravatz, D. E., Linn, S., Peurifoy, S. W., & Walker, R. (2008). Double bubble experiments in the three-torus. *Math Horizons, 15*(4), 18-21.

Charbonneau, L. (1996). From Euclid to Descartes: Algebra and its relation to geometry. In N. Bednarz, C. Kieran, & L. Lee (Eds.), *Approaches to algebra: Perspectives for research and teaching.* (pp. 15-37). Dordrecht: Kluwer.

Christou, C., Jones, K., Mousoulides, N., & Pittalis, M. (2006). Developing the 3DMath dynamic geometry software: Theoretical perspectives on design. *International Journal of Technology in Mathematics Education, 13*(4), 168-174.

Coxeter, H. S. M., & Greitzer, S. L. (1967). *Geometry revisited.* New York: Mathematical Association of America.

Department for Education and Employment (DfEE). (1998a). *Numeracy matters: The preliminary report of the Numeracy Task Force.* London: DfEE.

______. (1998b). *The implementation of the national numeracy strategy: The final report of the Numeracy Task Force.* London: DfEE.

______. (1999). *The national numeracy strategy: Framework for teaching mathematics from reception to year 6.* London: DfEE.

______. (2001). *Key stage 3 national strategy: Framework for teaching mathematics, Years 7, 8 and 9.* London: DfEE.

______. (1988). *Mathematics for ages 5 to 16.* London: DES.

Engineering Council. (1999). *Measuring the mathematics problem.* London: Engineering Council.

Giaquinto, M. (2007). *Visual thinking in mathematics.* Oxford: Oxford University Press.

Hoyles, C., & Jones, K. (1998). Proof in dynamic geometry contexts. *In*: C. Mammana and V. Villani (Eds), *Perspectives on the teaching of geometry for the 21st Century.* (pp. 121-128). Dordrecht: Kluwer.

Jones, K. (1998). Deductive and intuitive approaches to solving geometrical problems. In C. Mammana & V. Villani (Eds.), *Perspectives on the teaching of geometry for the 21st Century.* (pp. 78–83). Dordrecht: Kluwer.

Jones, K. (2000). Critical issues in the design of the geometry curriculum, In B. Barton (Ed.), *Readings in mathematics education* (pp. 75–91). Auckland, New Zealand: University of Auckland.

Jones, K. (2002). Issues in the teaching and learning of geometry. In L. Haggarty (Ed.), *Aspects of teaching secondary mathematics: Perspectives on practice* (pp. 121–139). London, UK: Routledge Falmer.

Jones, K., Mackrell, K., & Stevenson, I. (2009). Designing digital technologies and learning activities for different geometries. In C. Hoyles and J.-B. Lagrange (Eds.), *Mathematics education and technology: Rethinking the terrain* [The 17th ICMI Study]. New York: Springer. Chapter 4.

Jones, K., & Mooney, C. (2003). Making space for geometry in primary mathematics. In I. Thompson (Ed.), *Enhancing primary mathematics teaching.* (pp. 3–15). London: Open University Press.

Kvasz, L. (2005). Similarities and differences between the development of geometry and of algebra. In C. Cellucci & D. Gillies (Eds.), *Mathematical reasoning and heuristics* (pp. 25–47). London: King's College Publications.

Logothetti, D. & Coxeter, H. S. M. (1980). An interview with H. S. M. Coxeter, the king of geometry. *Two-Year College Mathematics Journal, 11*(1), 2–19.

London Mathematical Society (LMS). (1995). *Tackling the mathematics problem.* London: LMS.

Mammana, C., & Villani, V. (Eds.). (1998). *Perspectives on the teaching of geometry for the 21st Century.* Dordrecht: Kluwer.

National Mathematics Advisory Panel. (2008). *Foundations for success: Final report of the national mathematics advisory panel.* Washington, DC: U.S. Department of Education.

National Aeronautics and Space Administration (NASA) (2006). *NASA Achieves Breakthrough In Black Hole Simulation.* Website accessed 31 August 2009: http://www.nasa.gov/vision/universe/starsgalaxies/gwave.html

Office for Standards in Education (Ofsted). (2008). *Mathematics: understanding the score*. London: HMSO.

Qualifications and Curriculum Authority (QCA). (2004). *Interpreting the mathematics curriculum: Developing reasoning through algebra and geometry*. London: QCA.

______. (2005). *QCA futures: meeting the challenge*. London: QCA.

______. (2007). *The new secondary curriculum: What has changed and why?* London: QCA.

______. (2009). Website: *National curriculum: What has changed and why?* London: QCA. Website accessed 31 August 2009: http://curriculum.qcda.gov.uk/key-stages-3-and-4/developing-your-curriculum/what_has_changed_and_why/index.aspx

Royal Society. (2001). *Teaching and learning geometry 11-19*. London: Royal Society.

Usiskin, Z. (2004). *Should all students study a significant amount of algebra?* Keynote at the Dutch National Mathematics Days (NWD), 2004. [keynote also in *Nieuw Archief voor Wiskunde*, June 2004, 147-151]

Zachariades, T., Jones, K., Giannakoulias, E., Biza, I., Diacoumopoulos, D., & Souyoul, A. (2007). *Teaching calculus using dynamic geometric tools*. Southampton, UK: University of Southampton.

CHAPTER 15

LINKING GEOMETRY AND ALGEBRA THROUGH DYNAMIC AND INTERACTIVE GEOMETRY

Colette Laborde

THE NATURE OF MATHEMATICAL OBJECTS AND THE CRUCIAL ROLE OF REPRESENTATIONS

As so often stated since the time of ancient Greece, the nature of mathematical objects is by essence abstract. Mathematical objects are only indirectly accessible through representations (D'Amore, 2003, pp. 39–43; Duval, 2000) and this contributes to the paradoxical character of mathematical knowledge (Duval, ibid., p. 60):

> The only way of gaining access to them is using signs, words or symbols, expressions or drawings. But at the same time, mathematical objects must not be confused with the used semiotic representations.

Other researchers have stressed the importance of these semiotic systems under various names. Duval calls them *registers*. Bosch and Chevallard (1999) introduce the distinction between ostensive and non-ostensive ob-

Future Curricular Trends in School Algebra and Geometry: Proceedings of a Conference. pages 217–229

jects and argue that mathematicians have always considered their work as dealing with non-ostensive objects and that the treatment of ostensive objects (expressions, diagrams, formulas, graphical representations) plays just an auxiliary role for them. Moreno Armella (1999) claims that every cognitive activity is an action mediated by material or symbolic tools.

The construction process of a new concept involves the learning of how to represent it and to express it in different representation systems. The activity of solving mathematical problems, which is the essence of mathematics, is based on both an interplay between various registers and treatments within each register. Students encounter two kinds of difficulties:

- performing operations within a specific representation system. Duval calls treatments such operations internal to a system. Representing a point $(x, f(x))$ on the graph of a function when its x-coordinate is given, finding on the graph the points corresponding to local minimal or maximal values are examples of such treatments;
- linking various representations of the same concept: in particular when solving a problem students do not spontaneously resort to another representation system, if not asked to do it.

High school students, and even university students in mathematics who never question the coherence between calculus and geometry, experience these kinds of difficulties (Robert, 1998). For them, for example, there are two definitions of π, a geometrical one and a definition by means of an integral, but they do not consider a possible relation between both definitions.

Therefore, the main assumption guiding the studies that are presented in this paper is that the teaching of mathematics must help students learn how to adequately use various representations and to move between them if needed. Another assumption is that students learn from solving problems or non-routine tasks in which they cannot guess the expected solution but with which they have to engage by making use of their mathematical knowledge. Such tasks demand that students develop new solution means that may become the source of new knowledge. We refer here to the concept of a *didactical situation* as developed by Brousseau (1997). We attempted to design such tasks in which solving the task requires students to move between representations or to process within a specific representation system.

Chauvat (1999) distinguishes three modes of using a representation, according to the treatment of the link between the concept and its representation:

- an ideographic or illustrative mode in which the representation acts as label, only referring to an idea, to the concept: this is often the case for the use of a parabola for the students, the shape of parabola evokes for them a polynomial of second degree but no more than that;

- a nomographic mode in which one operates on the representation without relating to the concept the actions carried out on the representation: this is often the case for the use of geometric diagrams for students. For example, constructing a tangent line to a circle for them is often done by drawing a line touching the circle, only by using visual cues and not by using a geometrical property of the tangent line;
- an operational mode in which the link between the representation and the concept is used: for example finding whether a point with given coordinates belongs to the graph of a function by referring to the definition of the graph as the set of points $(x, f(x))$.

Chauvat deals with the link between a representation and the concept. We extend it to the link between two representations. In particular, we consider that a mode is *operational* if the use of a representation is carried out by taking account of the relationship with another representation or if this use is controlled by this relationship.

REPRESENTATION OF MATHEMATICAL OBJECTS IN DYNAMIC AND INTERACTIVE GEOMETRY ENVIRONMENTS

The essential feature of dynamic geometry environments lies in the very specific nature of their diagrams: these diagrams are variable and can be continuously modified while keeping their geometrical properties when dragged. However, other kinds of representations are also available in some dynamic geometry environments, such as graphs of functions or of families of parametric functions, algebraic expressions, but the new aspect of all these representations is that they are linked. While one of the representations is continuously changed, the other representations depending on the first one are also varying. Not only geometric elements can vary, but also numbers such as measures, coordinates, results of calculation and numbers in symbolic expressions of functions or in equations of curves.

Variation and variables constitute the essence of many mathematical objects. These notions can become tangible in dynamic geometry, which exteriorizes the variation of both geometric objects and numbers. In a way, dynamic geometry environments are a material counterpart of the mental process that Monge considered fundamental for students (Leçons données à l'Ecole Normale de l'an III, 1792):

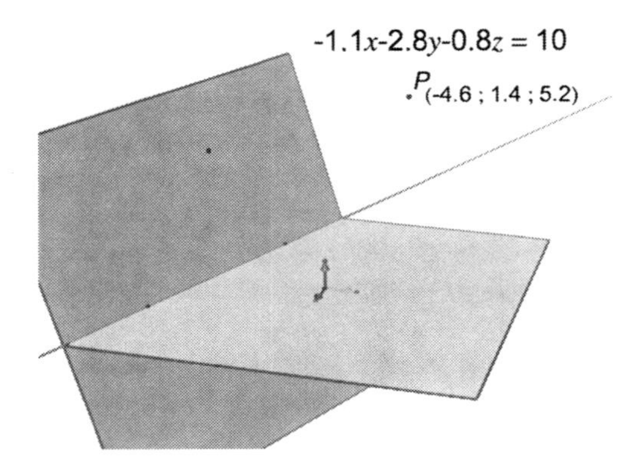

FIGURE 1. Plane and its equation

> Il faut que l'élève se mette en état, d'une part de pouvoir écrire tous les mouvements qu'il peut concevoir dans l'espace, et de l'autre, de se représenter perpétuellement dans l'espace le spectacle mouvant dont chacune des opérations analytiques est l'écriture.[1]

With their multiply-linked representations of mathematical objects "behaving mathematically," dynamic and interactive geometry environments constitute a place in which one can design tasks contributing to the flexibility between representations and, in particular, between geometrical representations and algebraic expressions of functions. Tasks in which students are asked to interpret or to produce given behaviors of these representations can take place only in dynamic environments. Let us illustrate this kind of task with an example in the dynamic 3D geometry environment Cabri.

Students are given a plane and its equation in Cabri 3D (Fig. 1). They can observe that, when moving a point defining the plane, the plane varies as well as its equation. A point *P* with displayed coordinates not belonging to the plane is given. Then the plane and all its points are hidden and students are asked to move point *P* in order to put it as close as possible to the hidden plane (Fig. 2). Of course students must solve the task with the plane remaining hidden.

Solving this task requires using the equation of the plane in absence of a spatial representation (Fig. 3). Algebra is the only way to control the position of the point with respect to the plane. This is exactly why the task is interesting for understanding that the value of the equation instantiated

[1]The student must be able, on the one hand to write all motions that he can conceive in the space, and on the other hand to perpetually imagine in the space, the moving spectacle that each analytical operation transcribes.

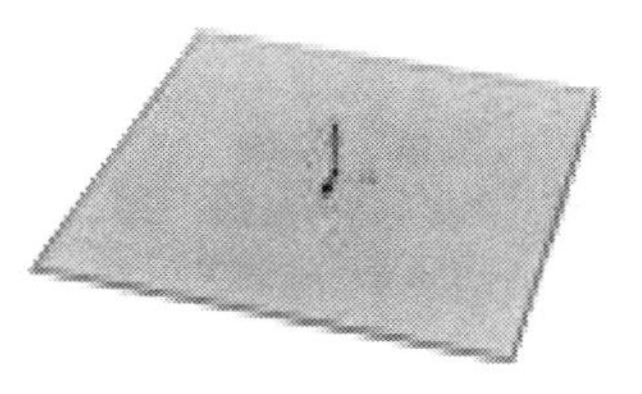

FIGURE 2. The plane is hidden until the expression equals 0.

with the value of the coordinates of P allows the students to know whether the point is on the plane, very close to the plane, or not. The mode of using the algebraic expression of the plane is operational, in the language of Chauvat. Obviously this task can only be given in a dynamic environment in which geometric and algebraic representations are linked.

Such tasks given in dynamic geometry environments can contribute to learning for two reasons:

1. the task itself and its demands
2. the feedback given by the environment to the students' actions and in particular, when dragging, the students may observe the effects

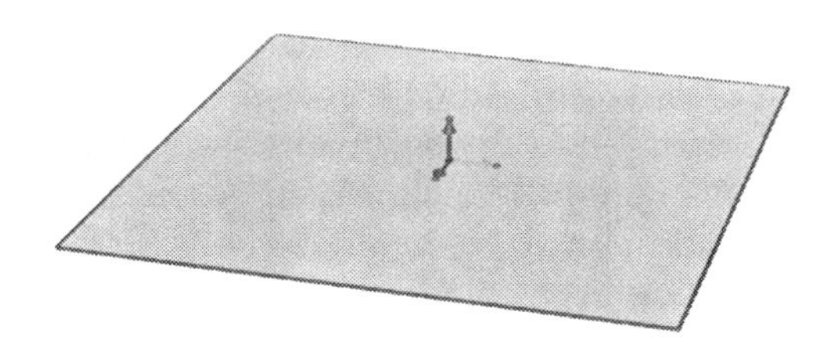

FIGURE 3. Point P is moved.

of their construction in the environment, whether it is preserved by drag mode or not, whether it behaves as expected or not.

Two examples of experimented tasks and teaching sequences making use of Cabri II Plus are developed in the next sections.

CONSTRUCTING THE MEANING OF THE NOTION OF GRAPH OF A FUNCTION

Consider the following teaching sequence for understanding the notion of graph of a function and learning to use it at grade 10. The concept of function is characterized by the large number of possible representations introduced in the teaching of mathematics, including tables of values, graphs, symbolic expressions, and tables of variations (in which the variations of the function as well as its local minimal and maximal values are displayed). The French national curriculum (grade 10, 15- to 16-year-old students) stresses the importance of learning how to move easily from one type of representation to another one. The NCTM focal points and connections recommend that:

> students at grade 8 translate among verbal, tabular, graphical and algebraic representations of functions and they describe how such aspects of a function as slope and *y*-intercept appear in different representations.

Each representation brings to the fore different features of the concept of function and is relevant for different uses and problems. For example, numerical problems are better solved with symbolic expressions whereas it is easier to have a global view of the behavior of the function on a graphical representation.

However, it seems that for students there is a lack of explicit relationship between function and graph (Dreyfus and Eisenberg, 1983). Difficulties of interpreting graphic information in terms of function are widely reported. Generally speaking students do not consider the graph of a function to be the representation of the relationship that exists between the variables, nor do they analyze their characteristics (Trigueros, 1996; Markovits et al., 1988; Schwarz and Hershkowitz, 1996; Cavanagh and Mitchelmore, 2000).

In summary, students have problem grasping the idea of a function as a relationship between variables (one depending on the other). They have a discrete view of a function, in which a function relates separate pairs of numbers, each number may be considered as an input giving another number as result: students consider that there is a relationship between numbers, but the relation is conceived separately for each pair. In any case, the relationship of dependency between the two variables is not visible in the graph, the graph remains a static representation of the ordered pair (x, y) and does not afford the meaning of dependency between the two variables and instead plays a symmetrical role.

The correspondence between a symbolic expression and a graphical representation of a function is based on the idea of representing a variable number by a variable point on an axis, i.e. of converting a numerical variation into a geometrical variation. Euler was the first one to write about this idea in a very explicit way (Euler, 1743):

> [...] Because, then, a unlimited straight line represents a variable quantity x, let's look for a method equally comfortable (useful) to represent any function of x geometrically.
>
> [...]Thus any function of x, geometrically interpreted in this manner, will correspond to a well defined line, straight or curve, the nature of which will depend on the nature of the function."

Euler proposed to represent a function as the curve obtained by all points M obtained in the following way: On an axis, each value of the variable x is represented by a point P. A segment $\overline{PM}$ perpendicular to the axis is constructed with length $f(x)$. When x varies, one considers all points M giving rise to a curve (Fig. 4).

Dynamic geometry can easily materialize the thought experiment of Euler by carrying out the same construction as the one he described and dragging the variable point P along the axis. The trajectory of point M can be obtained by using the Trace tool in Cabri (Fig. 5). In this way, the graph of the function can receive a twofold meaning, as the trajectory of the variable point M and as a curve. This dual meaning is absent for students most of the time because with paper and pencil only the curve aspect of the graph is at the fore. The curve is often drawn with paper and pencil by joining some points with a smooth line and students have no idea of the mathematical meaning of this curve. Such use of the graph is restricted to the

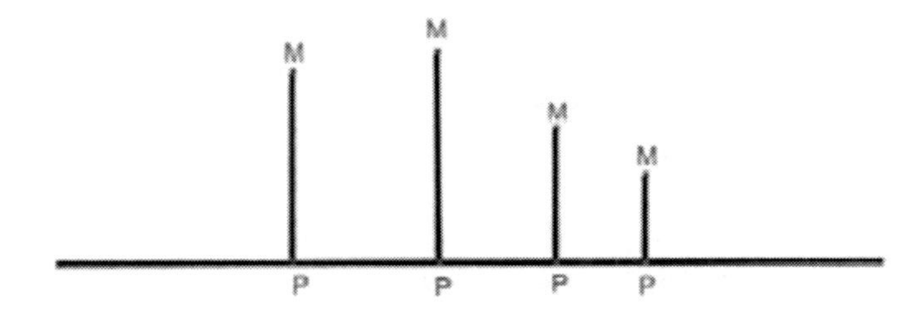

FIGURE 4. Generation of the graph of a function by Euler.

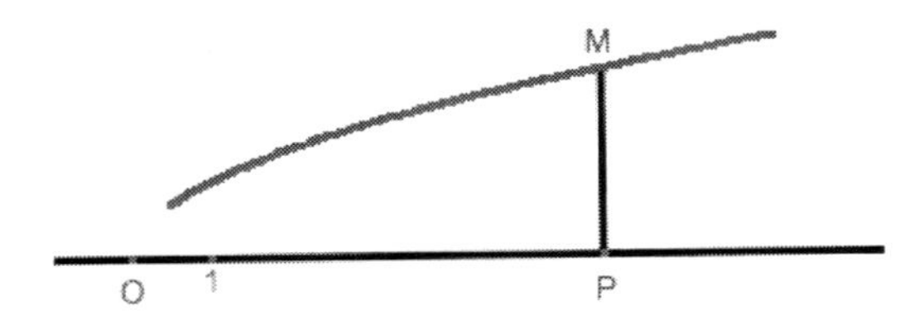

FIGURE 5. Euler's process carried out in Cabri

ideographic use, whereas dynamic geometry may restore such operational use as conceived by Euler.

Dynamic geometry introduces an asymmetry in the variables and allows students to become aware of the difference between independent and dependent variables defining a function. The independent variable can be dragged directly whereas the dependent variable can be moved only indirectly by dragging the independent variable. The theoretical distinction between the two variables is materialized in the dynamic geometry environment.

Based on these assumptions, a teaching sequence for introducing students to the notion of function and of graph of a function was designed and carried out in three classes at grade 10, two times in France and one time in Italy (Falcade, 2005; Falcade et al., 2007). The main idea was first to introduce students to geometric functions in Cabri through the expressive mediation of the dragging and trace tools in order to familiarize students to the notions of independent and dependent variables and of function, and only in a second step to move to numerical functions according to Euler's process. The sequence is described and some results are given in Falcade et al. (2007). Let us just mention that students internalized very well the difference between independent and dependent variables, even evoking the direct or indirect move of the variables when they were no longer working in Cabri. They also proved to be able to use in an operational way the graph to find the image and range of a function, as well as the preimage of subsets of the range of a numerical function.

DIFFERENTIAL EQUATIONS IN PRESERVICE TEACHER EDUCATION

We now turn to a situation for the learning of the link between the symbolic expression of a differential equation and the family of curves representing the function solutions of the equation at 4th university year for preservice secondary school mathematics teachers.

In the teaching of ordinary differential equations, the algebraic approach prevails internationally (Artigue, 1992). There are few tasks involving the use of graphical representations and most of them require a move from the algebraic representation to the graphical one. For example, students are required to first solve the differential equation and only then, to represent it. As a result, students have difficulties in interpreting graphical representations of slope field or solutions of differential equations and use them in an operational way. Furthermore, they still have difficulties even when the graphical approach is favored in the teaching they have seen (Rasmussen, 2001).

Moreno Gordillo (2006), in his doctoral dissertation work, investigated pre-service teachers' difficulties with first-order ordinary differential equations, although these 4th-year university students had already been familiar with these equations for several years (four or five years). Moreno Gordillo asked 56 students to interpret the picture of the vector field assigning the constant vector (1, 2) to each point of the plane (Fig. 6) as the slope field of a differential equation. Only two out of the 56 students succeeded in linking this representation to the differential equation $y' = 2$. Six other stu-

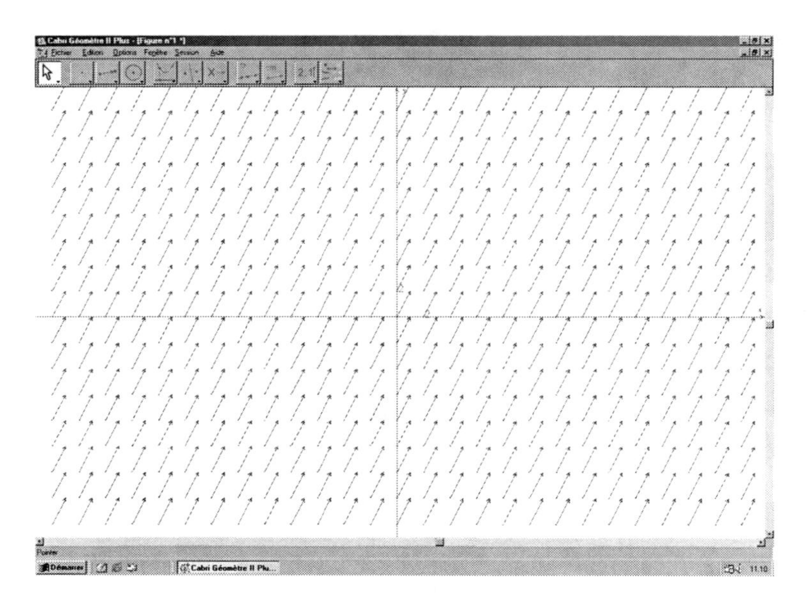

FIGURE 6. A vector field of a differential equation to be interpreted.

dents could state that the solutions are straight lines without giving their equations.

Moreno Gordillo gave the other following task to the same students.

> Let the DE be $y' = y^2 - x$. C is the curve representing a solution and passing through point $M(-2,1)$.
>
> Describe the behavior of C around point M. Give an approximate value of the y-coordinate of point N on C with x coordinate $x = -1.5$.

Only 4 students out of 56 described the local behavior of C. Among them, only 2 obtained the approximate value of the y-coordinate of N. 6 students (out of 56) tried to solve the differential equation.

Obviously those students had great difficulties in linking graphical and algebraic representations in an operational way.

Dynamic geometry environments may be the place for tasks requiring an operational mode of use of graphical representations of differential equations since they offer not only the possibility of constructing a variable curve of a family of parametric curves but also numerical and algebraic tools for exploring graphical representations. Inverse problems can also be given to students: instead of asking them to solve a differential equation, one can ask them to give the family of curves solutions and to find the differential equation. This type of task was given by Moreno Gordillo (2006) to the same students.

Students were asked to determine the differential equation of the family described only by a variable curve in Cabri II plus (Fig. 7).

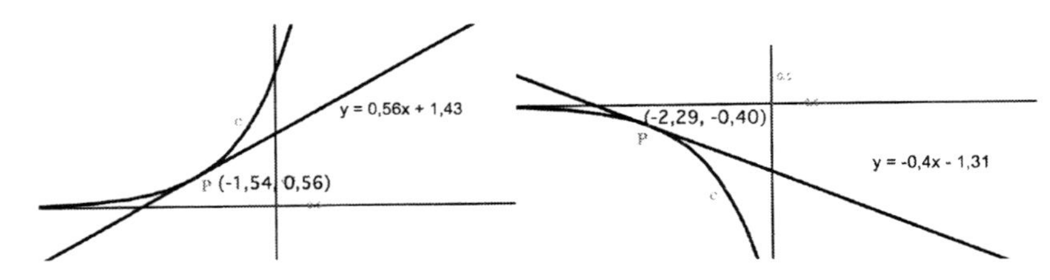

FIGURE 7. The variable curve in two different positions and a tangent line at a variable point P on the curve.

As this was the curve of an exponential function, three kinds of strategies were possible:

- recognizing the shape and guessing that the function is exponential, trying to find the parameters of the exponential by guessing, graphing and using the drag mode as a means of invalidation/validation;
- recognizing the invariance of the length of the subtangent (a geometrical strategy), expressing this invariant in an algebraic expres-

sion by giving the value y' to the slope of the tangent line; this strategy requires the expectation of a differential equation of the form $y' = f(x,y)$;
- displaying an equation of the tangent line at a variable point P of the variable curve, as well as the coordinates of P (an algebraic strategy); the equality of the slope of the tangent and of the y-coordinate of P is then inferred when dragging P.

All students developed the third strategy (algebraic) after some trials of the first strategy, but careful observation of their solving processes revealed deep difficulties and unexpected conceptions of a differential equation. Some students thought that the equation of the tangent line should be a differential equation and tried to adopt this strategy, which apparently failed, as the equation of the tangent line changed when P was dragged. Others thought that, when the equation of the tangent line was $y = ax + b$, the differential equation should be $y' = ay + b$. Some students were looking for a different differential equation for each position of the variable curve. Most of the students who noticed that for each P, $y_P = y'_P$ did not think that they had obtained a differential equation and did not know what to do with this expression. The teacher intervened by asking the students the following questions: 1) What does a differential equation express? 2) Can we express y' as a function of x and y? Only after such interventions, students realized that a differential equation could be conceived as a relationship between the slope of the tangent at a point and the coordinates of this point.

From these observations, it clearly appears that to large numbers of students:

- the notion of derivative in calculus is not related to the slope of the tangent line in coordinate geometry.
- a first-order differential equation is not viewed as a relation between the slope of the tangent line at each point of any curve of the family of the solutions and the coordinates of this point.

Students were not able to use this geometric interpretation although they were introduced to it in their studies. The relation between the family of curves and the differential equation students constructed from the teaching was a relation oriented from the equation to the curve that consisted of a two-step process—first, solving the given differential equation by means of a routinized procedure, and then, drawing the curve representing some solutions.

This task turned out to be a window on students' conceptions of differential equations. It also proved to be a learning situation, for when students were then faced with the task of finding the differential equation of a vari-

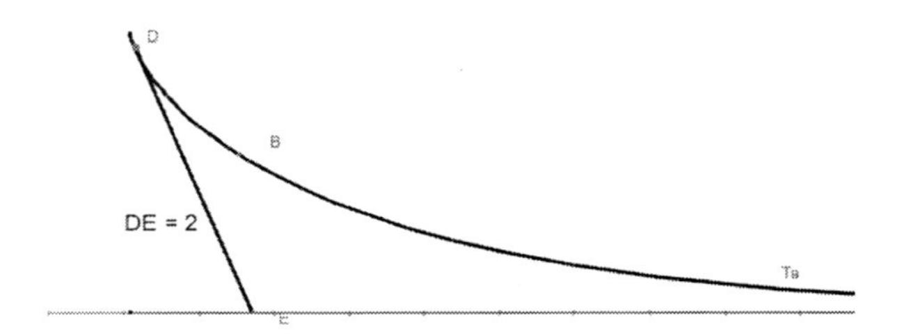

FIGURE 8. A variable tractrix.

able curve with tangent-line segment between the curve and the *x*-axis of constant length (tractrix) (Fig. 8), they were able to calculate the differential equation by expressing algebraically the invariance of the length of the tangent line segment.

CONCLUDING WORDS

With the availability of digital environments, several multiply-linked representations are possible, but simply facing students with such environments may not be sufficient to promote learning. New kinds of tasks made possible in these environments, in which mathematical questions must be solved using graphical representations, may contribute to the development of an operational use of these representations and therefore help for a richer conceptualization. The design of curricula should take advantage of such contributions.

REFERENCES

Artigue, M. (1992). Functions from an algebraic and graphic point of view: cognitive difficulties and teaching practices. In G. Harel & E. Dubinsky (Eds.), *The concept of function. aspects of epistemology and pedagogy* (pp. 109–132). MAA Notes No. 28. Washington, DC: Mathematical Association of America.

Bosch, M., & Chevallard, Y. (1999). La sensibilité de l'activité mathématique aux ostensifs *Recherches en didactique des mathématiques, 19.1,* 77–124.

Brousseau, G. (1997). *Theory of Didactical Situations in Mathematics.* (N. Balacheff, M. Cooper, R. Sutherland, & V. Warfield, Eds.). Dordrecht: Kluwer.

Cavanagh, M., & Mitchelmore, M. (2000). Student misconceptions in interpreting basic graphic calculator displays. In T. Nakahara & M. Koyama (Eds.), *Proceedings of the 24th Conference of the Group for the Psychology of Mathematics Education,* (Vol. 2, pp. 161–168) Hiroshima: Hiroshima University.

Chauvat, G. (1999). Courbes et fonctions au collège. *Petit x 51*, 23–44.

D'Amore, B. (2003). Le basi filosofische, pedagogiche, epistemologiche e concettuali della didattica della matematica, Bologna, Italy: Pitagora Editrice.

Dreyfus, T., & Eisenberg, T. (1983). The function concept in college students: Linearity, smoothness and periodicity. *Focus on Learning Problems in Mathematics, 5*, 119–132.

Duval, R. (2000). Basic issues for research in mathematics education. In T. Nakahara, & M. Koyama (Eds.), *Procedings of the 24th Conference of the International Group for the Psychology of Mathematics Education*, (Vol. 1, pp. 55–69) Hiroshima: Hiroshima University.

Euler, L. (1743). Introductio in analysin Infinitorum Tomus secundus Theoriam Linearum curvarum, Lousannae. English translation Introduction to the Analysis of the Infinite by John D. Blanton, Book 1, Springer Verlag (1988).

Falcade, R. (2005). *Théorie des situations, médiation sémiotique et discussions collectives dans des séquences d enseignement avec Cabri-géomètre pour la construction des notions de fonction et de graphe de fonction*, Doctoral dissertation. Grenoble, France: University Joseph Fourier.

Falcade, R., Laborde, C., & Mariotti, M. A. (2007). Approaching functions: Cabri tools as instruments of semiotic mediation, *Educational Studies in Mathematics, 66.3*, November 2007, 317–333

Markovits, Z., Eylon, B., & Bruckheimer, M. (1988). Difficulties students have with the function concept. In A. F. Coxford and A. P. Shulte (Eds.), *The Ideas of Algebra, K–12.* (pp. 43–60). Reston, VA: NCTM.

Monge, G. (1792). *Séances des Ecoles Normales recueillies par des sténographes et revues par les professeurs.* Première partie. Leçons Tome 1–5 Paris: L. Reynier s.d. (an III).

Moreno-Armella, L. (1999). Epistemologia ed Educazione matematica, *La Matemetica e la sua Didattica, 1*, 43–59.

Moreno-Gordillo, J. J. T. (2006). *Articulation des registres graphique et symbolique pour l'étude des équations différentielles avec Cabri géomètre*, Doctoral dissertation. Grenoble: University Joseph Fourier.

Rasmussen, C. (2001). New directions in differential equations: A framework for interpreting students understandings and difficulties. *Journal of Mathematical Behavior, 20*, 55–87.

Robert, A. (1998). Outils d'analyse des contenus mathématiques à enseigner au lycée et à l'université. *Recherches en didactique des mathématiques, 18*(2), 139–190.

Schwarz, B. B., & Hershkowitz, R. (1996). Effects of computerized tools on prototypes of the function concept. In L. Puig & A. Guttierez (Eds.), *Proceedings of the 20th PME Conference* (Vol. 4, pp. 267– 274). Valencia, Spain: University of Valencia.

Trigueros, M., Ursini, S., & Reyes, A. (1996). College Students' conceptions of variable. In L. Puig & A. Guttierez (Eds.), *Proceedings of the 20th Conference of the International Group for the Psychology of Mathematics Education* (Vol. 4, pp. 315–322). Valencia, Spain: University of Valencia.

CHAPTER 16

LINKING ALGEBRA AND GEOMETRY

The Dynamic Geometry Perspective

Nicholas Jackiw

In this paper, I explore the implications of dynamic geometry software on the effort to forge connections between school algebra and school geometry. Dynamic geometry itself has become a well-established technology paradigm in school mathematics over the past ten years; within the United States, the software with which I work, *The Geometer's Sketchpad* (Jackiw, 1991, 2001), has become one of the most widely used technology tools in both schools and pre-service education, and is the program that one broad survey of mathematics teachers, found "most valuable for students" (Becker, 1999)[1]. At present, in determining an effective curriculum followed by particular schools, educational technology of course plays a distant third to basal textbooks and state or provincial teaching expectations. But in the USA, and globally, *Sketchpad* is more widely adopted than any single basal text, and more generally positioned than any single state teaching framework. So one may say the potential exists to imagine broad impact on the

[1]The term "Dynamic Geometry" was itself originally coined to describe *Sketchpad (1991)*, and is a registered trademark of Key Curriculum Press. Since the time that *Sketchpad* and *Cabri* (Laborde et al., 1991) introduced interactive, geometric "dragging" to geometric construction software, many later programs have adopted similar manipulative and representational approaches.

Future Curricular Trends in School Algebra and Geometry:
Proceedings of a Conference. pages 231–241

algebra/geometry connection, brought about through the agency of dynamic technology.

Yet, what could this impact look like? The title of this paper alone—Dynamic Geometry—suggests that *Sketchpad* is a tool that speaks only to the geometry half of an algebra/geometry equation. What do draggable, rubbery triangles have to do with algebra? And the second half of the equation is no more certain, school algebra itself being a complicated, multi-faceted topic. Popular curricular approaches over the past two decades have focused on conceptions of school algebra as concerned centrally with functions (and featuring the graph of a function as a privileged representation); or as centrally about data (with a table of related values as a privileged representation); or, of course, as about symbolism and its mechanics (with expressions and symbolic notation as the most privileged representations). Many early software approaches to school algebra embraced this plurality through multiple simultaneous and linked representations, presenting the concept of a function through some symbolism representing its equation, through a table of its values, and through a graph of its projection on some coordinate system—all at the same time. And yet, if we pursue Lins' claim that students resist algebraic symbolism as "an alien entering the classroom," perhaps the way to help them become more comfortable with that alien is not immediately to encourage two new aliens, of different species, to follow it into the classroom as well! Students need to develop familiarity, if not fluency, with each representation before meaningful comparative activity can occur. The situation is not transparently better if we add dynamic geometry to the mix, for how does geometry relate to each of these representations, and to the underlying school algebra topics they variously construct?

ANALYTIC GEOMETRY: FOR AND AGAINST

The most obvious proposed linkage between algebra and geometry is, of course, the historical one: analytic geometry. In this approach, one represents geometric points by coordinate pairs, and geometric objects by various coordinate equations and relations, and then exploits accessible symbolic machinery both for the analysis of existing geometric configurations and the construction (or specification) of new ones. Analytic geometry occasioned the first major revolution in the mathematical development of geometry after Euclid, and already permeates some of the school treatment of elementary algebra.

Yet, even before considering a dynamic software perspective on analytic geometry, several limitations present themselves in considering it as a foundation for a pedagogic integration of algebra and geometry. First, lots of interesting geometry—triangles, for example—resist representation by straightforward analytic equations or as algebraic curves. Second, and related, a striking "grade-level mismatch" often exists between geometric expressitivity and analytic machinery. We introduce circles in kindergarten, but the quadratic equations that represent them analytically are not studied until high school. Third, there is the question of primacy—in analytic geometry, is the algebra or geometry primary? Through its vehicle, which sheds light on the other? Historically, analysis was deployed as a tool to solve problems in geometry, but in contemporary curricula, the reverse is the far more frequent case. Said differently, the presence of a *mathematical* bridge between two subject areas contributes little, *a priori*, to our understanding of the viability or shape of a possible *pedagogical* bridge between the same subjects.

These plausible limitations and risks become actual and real when analytic geometry's topical "bridge" is operationalized as the foundational principle of algebraic and geometric integration in dynamic software design. Many of them, for example, undermine the coherence of GeoGebra (Hohenwarter et al., 2005), an educational software package its designers describes as "dynamics mathematics software for schools that joins geometry, algebra, and calculus," or more simply by the equation: "GeoGebra = Geometry + Algebra" (Hohenwarter & Hohenwarter, 2008, p. 7). The software uses analytic geometry as the unifying perspective to bring these domains together, and the central concept of its user interface is that a *mathematical object* simultaneously appears in both a symbolic, "algebraic" form on the left and a graphical, "geometric" form on the right (see Figure 1).

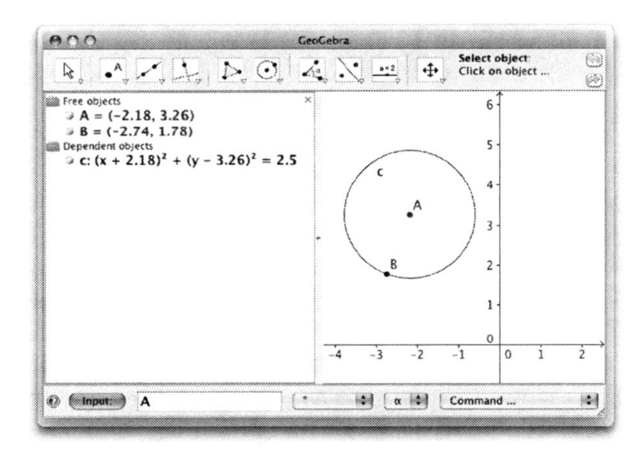

FIGURE 1. GeoGebra's "algebra + geometry" interface

At first blush, the idea appears natural and attractive: where we prefer the geometric characteristics of a shape, we focus on its visual image at right, but for the more analytic perspective, we turn to its equation or coordinates at left. Importantly, we are reminded of the mathematical equivalence—or even, identity—of these ideas or representations throughout. When we point to the representation of an object in one window, its corresponding representation highlights in the other as well; and when the program asks for one, it accepts the other in its place, because after all—the software constantly tells us—*mathematically the object and its equation are one and the same thing*. This is clearly the "strong" form of the analytic geometry bridge!

But no sooner does the traffic of school mathematics begin to move across this bridge than its robustness as a *total* metaphor—as a totalizing "solution" to problems of pedagogically linking algebra and geometry—begins to quaver. Figure 2 (top) adds a radius to Figure 1's circle, and shows that—within this paradigm of mathematical sameness—geometric segment a (the radius at the right) *is* the real number 1.58 (its depiction at left)! As educators, we might stumble on this before comprehending its internal logic. Presumably, an honest analytic equation for the segment is too obscure to show here. That is, the equation—and range restrictions—of the closed interval corresponding to a finite portion of an infinite line, expressed analytically in a formulation capable of encompassing the dynamic potential for variation of the figure at right, is a sufficiently unfriendly symbolic object from the school curriculum perspective—sufficiently "alien"—that in this case GeoGebra instead shows us the more *useful* value of a segment's length (rather than the more *consistent* "algebraic" expression of its definition). Thus a pedagogical decision in the software's design is motivated by the grade-level mismatch mentioned earlier. And while we can follow and understand the logic of the decision, we might equally regret its consequences for students. If the software implies that 1.58 *is* the segment radius of the given circle, then how on earth should they make sense of (1.58, 1.58)?

Figure 2 (bottom) shows a more fundamental example of the limitations of this type of "unification" of algebra and geometry. The previous construction is unchanged, but the unit spacing of the coordinate system's vertical axis has been adjusted. And the circle has stretched into a geometric ellipse—or has it? On the left, we can see it still has the same equation as before. And perhaps here our initial surprise gives way to a sense of mathematical relief: *of course* it's still a circle—we've just changed the coordinate system so it *looks* like an ellipse. But this does not quite work. For when we say a circle represents a constant distance from a point, we mean a *geometric* distance, of course: a physical dimension predating both people and their historically-recent invention of coordinate systems. And clearly the ellipse at right no longer conforms to a circle's *geometric* definition; instead, it preserves a constant analytic (or coordinate system) distance. And importantly,

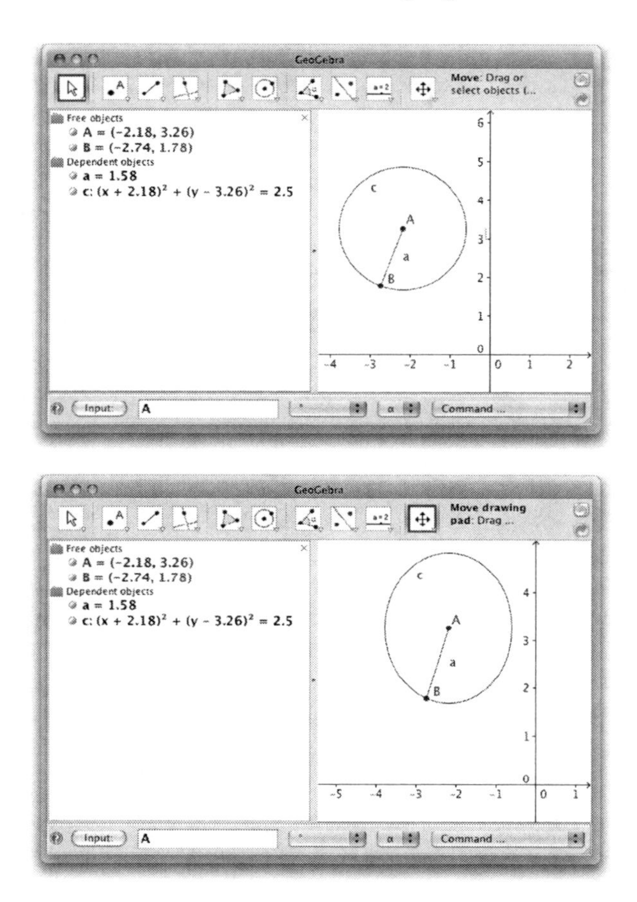

Figure 2. Pedagogic problems with the analytic geometry metaphor.
At the top, segments "become" numbers for want of more accessible analytic representations. At the bottom, "circles" become geometric ellipses for want of analytic consistency.

mathematically, geometric distance is *not the same thing* as analytic distance: where the former is a measure of length, the latter is a measure of ratios of lengths. (The former is dimensioned in inches, or cm, or light-years; the latter, as a ratio, is pure number.) Nor is this an abstruse objection, it's a foundational one. For if coordinate systems' axes measured geometric lengths, rather than analytic ones, simple analytic possibilities like $y = x^2$, which are dimensionally coherent (only) when x and y are unitless, would collapse into the geometric nonsense of $length = length^2$—that is to say: "lengths and areas are the same thing!"

Thus, the promised conceptual *unification* of geometry and algebra reveals itself as a false promise. The circle in the above pictures is no circle at all, geometrically. It's simply the plot of the implicit function $(x + 2.18)^2 + (y - 3.26)^2 = 2.5$ on a certain coordinate system. It is this particular form of quadratic function that is defined by a circle, not a geometric object, not a set of physical properties, and not a construction sequence. And that internal choice of representation arbitrarily limits how we interact with, and view, the "circle" in both the software's algebra and geometry windows. Algebraically, if general circles are *always* expressed as quadratic equations of this form, we cannot for example switch to simpler representations in which a circle might be modeled by a linear function or even a constant one (such as $f(\ ) = 1$). Likewise, if this specific circle is *defined* by the equation $(x + 2.18)^2 + (y - 3.26)^2 = 2.5$, then we cannot even execute the standard analytic geometry maneuver of choosing a more convenient origin to simplify the expression. (As you attempt to relocate GeoGebra's coordinate system origin—say to point A at the circle's center—your circle runs away from you, since it's implicitly defined as always being (2.18, –3.26) units away from that origin.) Thus, over-investing analytic geometry with the responsibility for unifying—rather than merely linking—algebra and geometry all too quickly undermines one's geometry, complicates one's algebra, and destabilizes the useful bridge between them that analytic geometry offered in the first place.[2]

DYNAMISM IN ALGEBRAIC THINKING

If there is no royal pedagogic road to algebra/geometry *unification*, we are back to the question of how best to link and connect them. Analytic geometry remains an effective way to link algebra and geometry once students have sufficient exposure to the representational and computational appa-

[2]My argument here is not that analytic geometry is mathematically inadequate or pedagogically inappropriate—of course it is neither. My claim is instead that we should not expect or demand it to unify algebra and geometry, in the literal sense of making one. At best analytic geometry loosely and multiply connects these two domains. It's instructive in this critique to compare *GeoGebra*'s performance tightly-coupling geometry to algebra, to the behavior of other dynamic software packages offering looser analytic couplings. In *Cabri II* and *Sketchpad*, changing the scale of the vertical coordinate system from square to non-square changes the measured equation of a circle into what we might for notational convenience call "the equation of an ellipse," without affecting the geometric circle at all. Similarly, moving the origin toward the center of the circle simplifies the measured equation, rather than displaces the circle itself. Users can view the same "geometric" circle through multiple different coordinate systems at the same time, and use multiple analytic descriptions (Cartesian, polar, etc.) simultaneously. Of course, this implies that there is no 1:1 equivalence between "an object and its equation"—no algebra/geometry unification—since multiple equations (or no equation) may describe a single geometric object, and since single equations may describe multiple objects (when plotted simultaneously on multiple coordinate systems) or none at all (if not plotted in the first place).

ratus that analysis pre-supposes. But it is not the only way. With its relatively high bar of accessibility within the spread of *school* mathematics, and its relatively limited topicality within the scope of school algebra and school geometry, it is perhaps not even the best way.

Returning to the perspective of dynamic mathematics technology, I would like to explore a conception of school algebra (from pre- and early-algebraic thinking at the elementary level, through the routine battery of secondary-level topics) as centrally concerned with numeric variation and co-variation. Even if few would claim that description as forming their own, pre-existing definition of algebra, most would recognize its substance in describing the various representational architectures used in teaching and learning about algebra. The graph depicts how one quantity changes (usually—and most usefully—in a continuous fashion)—according to change in another; the table or data column vector represents how multiple varying quantities somehow represent instances of "the same thing;" and of course, in symbolic forms, the promise and potential of quantitative variation becomes embodied by the variable x itself. Each of these representations places different relative emphasis on the importance of *number* compared to the importance of *variation*, but it is through their combination of both ideas—into numeric variation and covariation, into quantities that represent numbers that can, may, and do change—that they contribute to the development of algebraic thinking.

Following this line of thought, we arrive at a somewhat novel insight. Technology has only recently brought fundamentally *dynamic* conceptions to triangles, measurement, and the topics of school geometry into the classroom—this is dynamic geometry. Algebra, on the other hand, has *always* been about what we might usefully call *dynamic number.*[3] And so we might reasonably ask: between a dynamic number conception of algebra, and a dynamic geometry conception of geometry, can we expect to find—or build—strong interdisciplinary linkages on the principle of dynamism itself? And if so, what might they look like?

Three strongly-worded advisories—whether warnings or invitations are not yet clear—confront the traveler setting out on such an intellectual exploration. First, our customary representations of algebra range from the somewhat impoverished to the thoroughly inadequate in their invocation of *dynamic* variation. (Our present-day symbolic notation exists at the end of a long historical trajectory whose evolutionary pressures have sought explicitly to limit the depiction of a quantity's *actual variation*, and to detemporalize the dynamic process of a quantity's *actually varying*). Second, our

[3]Tatha (1981), describing Gattegno's work, puts it like this: "[Where] geometry is an awareness of imagery, [s]uch awareness arises from a dynamic process of the mind, and any formalisation of such awareness is an algebraic activity.... Thus, *algebra is an awareness of dynamics* [...]." (p. 6, original stress)

visual cognition of dynamics is vastly superior to our numeric or symbolic cognition—whether of dynamic or of static cognizance. (This too, is an evolutionary point. We've co-evolved with our present mathematical symbol and number systems for only a few thousand years, where our visual cognition—our real-time visual systems for recognition, inference, and extrapolation—has been improving many hundreds of times as long.) Finally, new technologies (such as dynamic geometry), in exploiting this strong visual cognition, propose fundamentally new representational infrastructures for mathematics. All of this suggests we may not recognize this bridge as we cross it, or—having crossed—we may no longer recognize the landscape at which we've arrived, nor the one from which we came.

Fortunately, the history of dynamic representation technologies gives us some examples and preliminary insight into how such linkages may appear. An early instance may be found in Diderot's *Encyclopédie* (1751), in which D'Alembert's entry for "equation" devotes much of its space to the blueprint of a fantastic dynamic machine capable of embodying in its internal motions (or, as we would say: "capable of plotting") polynomial functions of arbitrary degree (Figure 3). At the end of the entry, we learn that due to the constraints of friction, the machine is impossible to realize, physically, for curves of degree greater than two! The idea of the machine, in other words, is more important than the machine itself, for it is the idea of the machine which is as mathematically comprehensive and general as the idea of the polynomial equation itself, while simultaneously being as concrete and literal and superficially obvious—obvious on its surface—as a child's plaything. It took 250 years for these "universal machines" to become more than pedagogic thought-experiments, and to become actually and physically palpable within the idealized, friction-free universe of dynamic geometry. But mathematical thinking in the Enlightenment is filled with them.[4]

Goldenberg, Lewis and O'Keefe (1992) describe a more recent technological example, the *Dynagraph*, stemming from research into then-new graphing software:

> [R]esearch showed that, faced with an expression like $f(x) = ax^2 + bx + c \ldots$, students had little idea of what made x so special. It was called "the variable," but they never varied it! On the contrary, what they varied were a, b, and c, which were called parameters or, even more mysteriously, constants! (Goldenberg, Scher, and Feurzeig, 2008, p.77)

In the *Dynagraph* experiment, students use interactive software to manipulate—vary—a variable defined in a geometric instantiation of its domain

[4]For a detailed argument about the pedagogic connection between Enlightenment mechanistic philosophy and dynamic geometry, see Jackiw (2006). I'm indebted to Jean-Marie Laborde, and through him to Michel Carral and Roger Cuppens, for first introducing me to D'Alembert's marvelous machine.

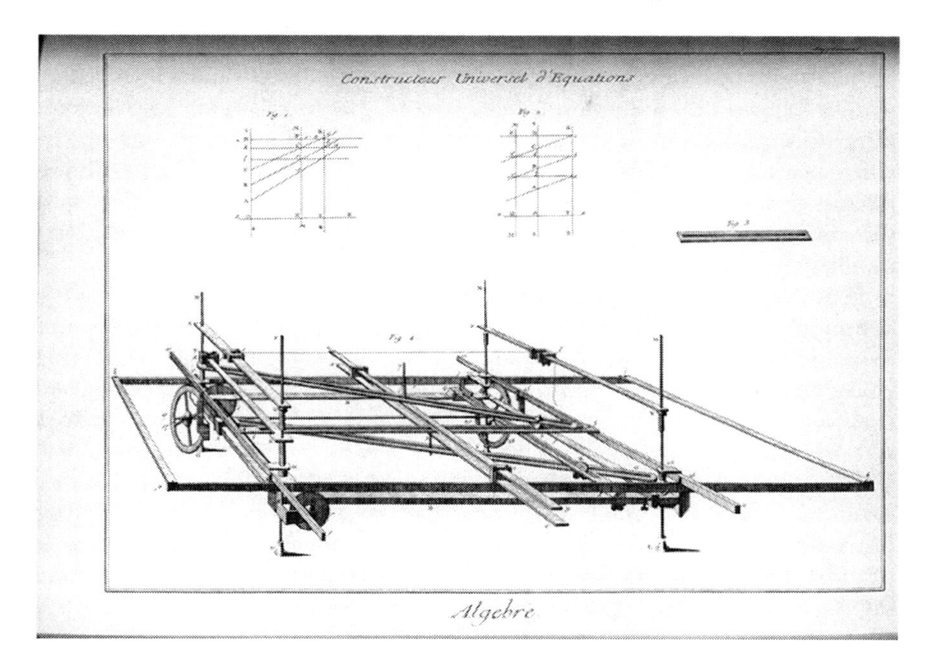

FIGURE 3. Le Constructeur Universel d'Équations, of D'Alembert

(a line or plane), while observing, predicting, and determining the resulting motion of a dependent variable depicted geometrically in some parallel range. Describing the motivation behind the *Dynagraph's* explicit link to geometry, the authors write: "what led to a consideration of geometric objects was the observation that students' concepts of functions were extremely narrow. The team began to think of ways to make accessible the idea of functions on nonnumeric spaces, including geometric spaces." (p. 77)

Moving into the realm of applications of existing dynamic geometry software, the first systematic reconstruction of Descartes' original analytic bridge within such software—to my knowledge—was the work of two graduate students working with *Sketchpad* at the University of Georgia in the early 1990's—long before *Sketchpad* or *Cabri* had analytic functionality "built-in" (see Lin and Hsieh, 1993). As with D'Alembert's machine, a student's reconstruction of Descartes within dynamic geometry is not simply an exercise in recreating historically important examples or illustrations. Since a dynamic geometry construction establishes an equivalence class of possible constructions, these recreations function more generally and more universally than their precedents, and become tools—student-built, and therefore, student-invested—for expressing, transforming, and solving broad ranges of similar problems to the one originally modeled. Moving beyond a focus on the equations of curves and graphs of functions, Chanan, Bergofsky, and Bennett (2002) provided perhaps the earliest attempt to construct a systematic secondary-level algebra curriculum—from arithmetic and simple expressions through the introduction of slope to first- and second-degree equations, factoring, root-finding, and so on—around dynamic geometry activity.

From this spectrum of preliminary examples, we can begin to identify common conceptual and procedural links between dynamic geometry and dynamic number—between a spatial/temporal and a symbolic/algebraic encoding of changing quantity. Indeed, these correspondences reveal themselves in almost any mathematical activity within *Sketchpad* that does not arbitrarily (or doctrinally) limit itself *only* to the software's geometric vocabulary, or only to its algebraic vocabulary. The role of *given points* in a geometric construction mirror those of *independent parameters* in an analytic derivation—the original inputs to one's mathematical process, manipulable but not determined. *Construction* and *transformation* give geometric meaning to *arithmetic* and *algebraic manipulation*. *Dragging* and *animation*—the core activities of dynamic geometry—stand for *varying the variables* in algebraic reasoning and problem solving. A *locus* can generalize the definition of a constructed point just as a *function* can generalize one value's defined dependence on another. Just as these ideas represent correspondences across two domains, we encounter explicit moves between them. From the perspective of easily-linked algebra and geometry, perhaps *measurement* is less strictly about magnitude, dimensionality, and error than it may stand as the simple mathematical act of *turning shape into number*. Conversely, plotting—whether a value on an axis, an ordered pair on the plane, or an equation as a curve or surface—appears as a recurring mathematical tool for *turning number into shape*.

These conceits—operation, variation, and generalization, harmonizing shape and number equally—form the basis of a vision of mathematics education in which geometry and algebra sit well together; one where learners can be at home in, and move comfortably between, both. Dynamic geometry software helps bring that vision into focus, first by establishing a palpable and operational premise (concept and image) of dynamic mathematical variation; second, by establishing on that central idea formal mathematical languages of geometry and of algebra; and third, by making explicit the connections, correspondences, and conversions between those languages. If we pursue this vision, however—if we pursue dynamic geometry as a tool for understanding algebra in just the way Descartes proposes algebra as a tool for understanding geometry—we must acknowledge we are adopting a fundamentally new representational infrastructure for our school mathematics. And, as Kaput (1998) often reminded us, changing our representational infrastructure will change not only our curriculum, but eventually, our mathematics itself.

REFERENCES

Becker, H. J., Ravitz, J. L., & Wong, Y. (1999). *Teaching, learning and computing 1998 national survey*. University of California, Irvine: Center for Research on Information Technology and Organizations.

Chanan, S., Bergofsky, E., & Bennett, D. (2002). *Exploring algebra*. Emeryville: Key Curriculum Press

Diderot, D. & le Rond d'Alembert, J. (Eds.), (1751). *Encyclopédie*. Paris, France: Andre Le Breton, Michel-Antonine David, Laurent Durand, & Antoine-Claude Briasson.

Goldenberg, E., Lewis, P., & O' Keefe, J. (1992). Dynamic representation and the development of a process understanding of function. In G. Harel & E. Dubinsky (Eds.), *The concept of function: Aspects of epistemology and pedagogy*. MAA Notes Number 25. Washington DC: Mathematical Association of America.

Goldenberg, E. P., Scher, D., & Feurzeig, N. (2008). What lies behind dynamic interactive geometry software?, In G. W. Blume & M. K. Heid (Eds.), *Research on technology and the teaching and learning of mathematics: Volume 2—Cases and perspectives*. (pp. 53–88). Charlotte: Information Age.

Hohenwarter, M. et al. (2005). GeoGebra [computer software]. Tallahassee, FL: GeoGebra Inc.

Hohenwarter, M., & Hohenwarter, J. (2008). *Introduction to GeoGebra*. Judith Hohenwarter/Lulu Publications.

Jackiw, N. (1991, 2001). The Geometer's Sketchpad (Versions 1-4) [computer software]. Berkeley, CA: Key Curriculum Press.

Jackiw, N. (2006). Mechanism and magic in the psychology of Dynamic Geometry. In N. Sinclair, D. Pimm, & W. Higginson (Eds.), *Mathematics and the aesthetic: New approaches to an ancient affinity* (pp. 145–159). Heidelberg: Springer.

Kaput, J. J. (1998). Representations, inscriptions, descriptions and learning: A kaleidoscope of windows. *Journal of Mathematical Behavior, 17*(2), 265–281.

Laborde, J.-M. et al. (1991, 1995). Cabri and Cabri II [computer software]. Grenoble, France: IMAG-CNRS.

Lin, P. -P., & Hsieh, C. -J. (1993). Parameter effects and solving linear equations in dynamic, linked, multiple representation environments. *The Mathematics Educator, 4*(1), 25–33.

Tatha, D. (1981). About geometry. *For the Learning of Mathematics, 1*(1), 2–9.

CHAPTER 17

LINKING ALGEBRA AND GEOMETRY IN THE INTERACTIVE MATHEMATICS PROGRAM

Diane Resek

The authors of the *Interactive Mathematics Program* (IMP) did not set out to write an integrated curriculum, per se; rather we wanted to organize the curriculum so learners would become invested in the subject matter. At the same time, we felt it was imperative that our curriculum would develop students' abilities to solve non-routine problems. In that respect we were responding to the ideas of that time, the late 1980's. That vision of problem solving is expressed in the following quote from *Everybody Counts* (National Research Council, 1989):

> From the accountant who explores the consequences of changes in tax law to the engineer who designs a new aircraft, the practitioner of mathematics in the computer age is more likely to solve equations by computer-generated graphs and calculations than by manual algebraic manipulations. Mathematics today involves far more than calculation; clarification of the problem, deduction of consequences, formulation of alternatives, and development of appropriate tools are as much a part of the modern mathematician's craft as are solving equations or providing answers (p. 5).

Future Curricular Trends in School Algebra and Geometry: Proceedings of a Conference. pages 243–249

Our two goals of engaging students and of creating authentic problem solvers led us to create an organizational scheme built around large, rich problems. We ended up with five such problems for each of the four years of the program. We wanted to teach all of the mathematics skills and concepts needed in secondary school in the context of those problems. One effect of this organizational scheme was that the curriculum became integrated. Not only was algebra integrated with geometry, but probability, statistics, and logic were also part of the mix.

MEADOWS OR MALLS? UNIT

The linking of geometry and algebra in IMP is best seen through the examination of one of IMP's twenty problem-based units. The examination can also show how concepts and skills are learned in the context of a large problem. The unit I have chosen as an illustration is from the third year of IMP. Aspects of algebra and geometry complement each other in various parts of the unit.

The unit is called *Meadows or Malls?*, and it opens by introducing the unit problem. Three separate gifts of land are made to a city, and the local environmental group and the business community each want to use the land for their own purposes: recreation vs. development. The city council comes up with a compromise, involving conditions such as the amount of one of the pieces of land that is used for recreation must be less than the amount of land on the other two pieces that are used for development. The compromise can be expressed in equations and inequalities using six variables: one variable for the part of each of the three pieces of land used for development and one variable for the part of each piece that is used for recreation. There are different costs to the city for improving the land so that it can be used for different purposes. The per-acre cost associated with each of the six variables is presented to students and they are asked to pretend that they are a team of consultants reporting to the City Manager. The Manager has asked them to find a way to satisfy each part of the compromise at a minimal cost to the city. The problem is a standard linear programming problem in six variables. The students learned how to solve linear programming problems in two variables the year before.

After reading the problem, students work in groups to get some values for the six variables that satisfy the compromise, and they try to keep the cost to the city low. They are getting invested in the problem and developing some intuition about it. Next they review linear programming in two variables. The previous year they had solved the problems visually or geometrically without formulating an algebraic procedure. To solve linear programming problems with four or more variables, they need to move from

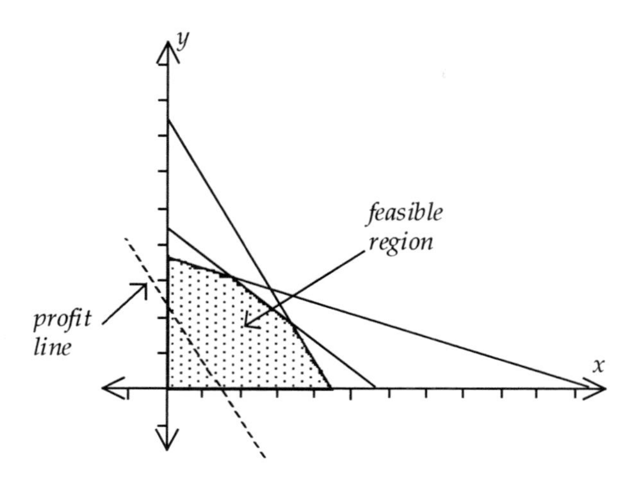

FIGURE 1. A diagram for a linear programming problem.

the geometric representation to an algebraic procedure, since graphs are not possible.

A linear programming problem in two variables involves some linear inequalities that must be satisfied and a linear expression to be maximized, such as a profit, or a linear expression to be minimized, such as cost. Typically, the points satisfying the linear inequalities lie inside a polygon as seen in the dotted area in the example shown in Figure 1 below. The set of points that satisfy all the inequalities is called the feasible region.

The solid lines are the boundaries of the inequalities. In this case, the dotted line represents all points giving a set profit, say $1,000. The points giving a higher profit would lie on a parallel line to the dotted one, since the slopes of the two lines would be the same. Lines of points giving different profits or costs are always parallel in a linear programming problem, since the same linear expression is being set equal to different numbers. In the case shown in Figure 1, if the profit is higher the y-intercept will be greater. The expression will be maximized at the point in the polygon on the parallel line with the greatest y-intercept. Students picture a sequence of parallel lines with increasing y-intercepts to find that last point. The point is always at a corner or sometimes along a side of the polygon. Thus to maximize profits, students could simply check the profit at each corner. So an algebraic procedure for finding the pair of points maximizing profit would be to check the profit at the points on the intersection of each pair of lines, and select the largest profit. Finding the minimal cost has a parallel procedure. Students formalize this algebraic principle in the beginning of the *Meadows or Malls?* unit. The algebraic procedure was not discussed formally the year before since we wanted to highlight the graphical reasoning or the geometry of the situation.

Next, students look at linear programming in three variables. They do not know how to graph in three variables, but they realize that it might help to be able to graph since they could then visualize the problem as they did when they worked with two variables. They proceed to learn about graphing equations and inequalities in three variables, and see that the feasible region in a linear programming problem in three variables is a polyhedron. The graph of an equation giving a set profit is a plane. Different profits yield parallel planes. As the profits rise, students see a sequence of parallel planes cutting through the polyhedron. The point in the feasible region that would maximize profits would be a corner point, or if the profit planes were parallel to a side of the polyhedron, any point on that side would give the same maximal profit. To find the maximal profit, or minimal cost, students need to simply check the corner points where three planes intersect. That is, they can consider the profits or costs at the solution points, if any, of triples of equations. The general solution to the linear programming problem is the greatest or least of those points.

When students worked on the linear programming unit in year 2, they invented procedures for solving two equations with two unknowns. Their inventions were variations on the substitution method. In this year 3 unit, they are told that in preparation for an easier way to solve systems of equations (matrices on their calculators), they need to learn the elimination of variables method for solving systems. They are led to discover this method by a sequence of word problems, such as: If 3 apples and 2 bananas cost $1.75 and 3 apples plus 4 bananas cost $2.75, how much do apples and bananas sell for?

After some practice solving systems of three equations, students return to the unit problems with six variables. They are asked to generalize their method for solving linear programming problems in three variables to linear programming problems in n variables. The procedure they devise typically involves listing the solutions to each set of n equations in the n unknowns, (the corner points), and then checking the profit or cost at those points. They have proved that this strategy works when $n = 2$ or $n = 3$ through the geometry of the situation. The teacher needs to assure them that there is an algebraic proof that their procedure is correct when n is greater than 3, but they will have to wait to see the proof after they have taken more mathematics.

At this point students try their procedure on a linear programming problem in four variables and realized that solving four equations with four unknowns is daunting. They are told that there is an easier way of solving systems of equations and they begin to work with matrices. They learn to multiply matrices, the connection of matrices to equations, and some matrix algebra. Students see they can multiply by the inverse of a matrix, (if it exists), to solve a system of equations. However, finding the inverse is, in general, as hard as solving the system. Finally, they are introduced to the matrix functions on their graphing calculators, which makes solving six equations in six unknowns straightforward. At last, they solve the unit problem and write an explanation of their work for the fictional City Manager.

ROLE OF REFLECTION

Often, for students to understand connections, the connection must be made explicit. Thus, for students to understand the interaction of algebra with geometry it is helpful to reflect on those connections. For this reason, students were asked on several occasions to write about the connections between the geometry and algebra that they had encountered in the unit. One example is shown in Figure 2.

Homework 22: Three Variables, Continued

You know that the graph of a linear equation in three variables is a plane in 3-space. You also know that the intersection of three planes can be

- nothing
- a single point
- infinitely many points (either a line or a plane)

Each of the problems below gives a system of three linear equations, so their graphs are three planes. For each system, state which of the three types of intersection the graphs have, and justify your answer. If the intersection is a single point, find that point.

1. $a - b + 2c = 2$
   $2a + 2b - c = -3$
   $3a + b + c = 4$
2. $r + s + t = 2$
   $2r + 2s + 2t = 4$
   $3r + 3s + 3t = 6$
3. $u - v + w = 2$
   $u + v + w = 6$
   $u + v + 2w = 9$
4. $2x = 6$
   $3y + z = -7$
   $6x + 6y + 2z = 4$

FIGURE 2. A three-variable problem.

Homework 36: Beginning Portfolios—Part I

This unit has involved several closely related ideas:

- Graphing linear equations in three variables
- Solving systems of linear equations in three variables
- Finding intersections of planes in 3-space

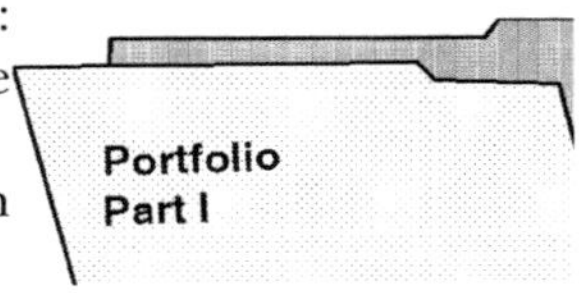

Portfolio
Part I

1. Summarize how these ideas are related. In particular, focus on the following two questions and how they are connected:
   - What are the possible results from solving a system of three linear equations in three variables?
   - What are the possible results of the intersection of three planes in 3-space?
2. Select activities from the unit that were important in developing your understanding of the ideas you discussed in Question 1, and explain why you made the selections you did.

FIGURE 3. Beginning a portfolio.

In IMP, students are asked to prepare a unit portfolio at the end of each unit. Preparing the portfolio requires students to reflect on connections made. Figure 3 shows one part of the portfolio where students are asked to reflect on some connections between algebra and geometry in the unit.

PROBLEM SOLVING AND AN INTEGRATED APPROACH

The previous sections have laid out the mathematics that students see in one unit. An argument can be made that this integrated, problem-centered approach allows students to develop stronger problem-solving skills than they would in a more compartmentalized curriculum. One can see those skills at work in Meadows or Malls? when students use the geometric justification in two- and three-variable linear programming problems to conjecture that an analogous algebraic procedure holds for higher dimensions. Students are unable to prove their conjecture is valid algebraically, but they can see an avenue they would pursue if they knew more mathematics. Going back and forth between various representations to formulate fruitful problem-solving approaches is a time-honored tool of mathematicians. If students are to become the type of problem solvers described in the quote from *Everybody Counts* at the beginning of this paper, they should become accustomed to looking at problems from different points of view.

Later, in a unit from the fourth year of IMP, *As the Cube Turns,* students move more frequently between the geometric and algebraic representations. The large problem of the unit is to program a graphing calculator to show a cube rotating in 3-space. This is a geometric problem but students must use algebraic expressions to program the calculator. Some of the algebra needed is discovered by considering the graphs of functions and the geometry of the situations.

CONFERENCE PERSPECTIVE

I was asked to characterize how linking geometry with algebra has changed in schools in my areas over the last decade. The practice of making connections between areas in the curriculum has lapsed in California, which is my area of the country, in the last decade. Schools in California have been under extreme pressure to teach to the state standards. Those standards emphasize the mastery of isolated procedures. For this reason, most secondary schools that used integrated curricula in California ten or fifteen years ago have gone back to other curricula in the last decade which are easier to align to the state standards.

I was then asked to reflect on what I would like to see happen in California schools. I would like us all to go back to the spirit and the vision of the late 1980's as seen in *Everybody Counts* and the *NCTM Standards.* In particular, I would like schools to focus on engaging students and helping them become powerful problem solvers.

REFERENCES

Alper, L., Fendel, D. M., Fraser, S., & Resek, D. (2003). *Interactive Mathematics Program: Year 2.* Emeryville: Key Curriculum Press.

______. (2003). *Interactive Mathematics Program: Year 3.* Emeryville: Key Curriculum Press.

National Research Council. (1989). *Everybody Counts - A Report to the Nation on the Future of Mathematics Education.* Washington, D.C.: National Academy Press.

PART V

PANEL OF PRACTITIONERS AND CURRICULUM DEVELOPERS

CHAPTER 18

MAKING FUTURE TRENDS REALITIES IN US CLASSROOMS

Diane J. Briars

The conference sessions provided a rich collection of ideas and technologies for school algebra and geometry and a dynamic vision for algebra and geometry instruction, especially at the high school level. A recurring theme of the high school sessions was the gap between this vision and the reality of today's classrooms. Despite the importance of topics such as three-dimensional geometry and the learning opportunities afforded by new technologies such as hand-held computer algebra systems (CAS) and dynamic 3-D geometric software, few of the ideas and newer technologies are used regularly in U.S. high schools. Although older technologies such as *Geometer's Sketchpad* and graphing calculators are used regularly in many US secondary classrooms, these newer technologies are not yet making headway. From my experience as a district mathematics coordinator, a number of issues and challenges need to be addressed for widespread implementation of the technologies presented at this conference. At the same time, the current environment affords opportunities that developers might consider to support implementation of their innovations.

Future Curricular Trends in School Algebra and Geometry: Proceedings of a Conference. pages 253–258

IMPLEMENTATION ISSUES AND CHALLENGES

Schools and Districts versus Teachers as Units of Change

In their sessions, the majority of developers described specific efforts to prepare teachers to implement their innovations, either working with small groups of preservice or inservice teachers in various professional development projects. The expectation was that individual teachers would then use these innovations in their classrooms, and even encourage their colleagues to do so as well. Unfortunately, the majority of teachers failed to consistently use the new technologies, even after considerable professional development regarding their use.

Upon further reflection, it is not surprising that this model—providing professional development to small groups of teachers—is not successful. The decisions regarding curriculum, instruction, and assessment that implementing these technologies involve, and the on-going support needed for teachers to use these technologies effectively, are beyond the purview of individual teachers. Schools, or school districts, rather than individual teachers, need to be the unit of change.

Curriculum

The potential impact of technology on mathematics curricula has long been recognized. Twenty years ago, NCTM (1989) summarized it concisely: Some mathematics becomes less important because technology replaces it, some becomes more important because technology requires it, and some becomes possible because technology allows it. Or, as Kaye Stacey described in her presentation, technology changes the curriculum value each topic, that is, its epistemic value, pedagogical value, and pragmatic value. Clearly, decisions about the relative value of curriculum topics need to made at the school, district and/or state level, rather than by individual teachers.

Making substantial changes in the high school curriculum, especially regarding the use of technological tools vs. paper-and-pencil symbol manipulation, poses a special challenge. Schools and school districts are accountable for preparing students for the workplace or additional schooling. Thus, curricular decisions about technology use vs. paper and pencil skills cannot be made independently of college and university requirements. College and university mathematics placement tests have significant effects on students. These tests are currently one of the major areas of mismatch between 9-12 and 13-16 education. And while various national groups provide standards documents about what students should know to be college and work-ready (Achieve, College Board, ACT), these documents are primarily directed to preK-12 institutions, with little, if any, effect on higher education institutions. In short, secondary schools cannot take the lead regarding the use of CAS systems to *replace* traditionally expected symbol manipulation skills, due to their responsibilities to prepare students for their future endeavors, which, for college-bound students, often includes taking college and university placement tests with traditional content.

Perhaps in recognition of this articulation issue between secondary and higher education and the expectations regarding paper-and-pencil symbol manipulation, the conference sessions about using CAS in the US high schools (Cuoco, Heid), illustrated how CAS can be used to support, rather than replace, acquisition of paper-and-pencil symbol manipulation skills that are typically required of high school students. Using CAS as a platform for experimenting with algebraic expressions and equations (Cuoco) and to develop symbol sense (Heid) are certainly valuable ideas to incorporate into the high school curriculum. However, we also need to take the next steps and use the capabilities of CAS to reduce computational requirements, especially for struggling students. A next step in this direction might be reconsidering *when* students are expected to demonstrate specific symbol manipulation skills, rather than the *amount* that is expected. Twenty years ago, Heid (1988) demonstrated that using technology to resequence instruction in skills and concepts in a calculus class, i.e., using technology to focus on concept acquisition during the majority of the course, then introducing skills instruction during the later part of the course instead of integrating both throughout, resulted in students acquiring better conceptual understanding, with virtually comparable skills. Perhaps it is time to try a similar resequencing in high school courses.

Instructional Resources

A second major challenge in implementing new technology is the availability of lessons using new technologies to teach specific content. Typically, individual teachers are not in a position to develop high quality lessons that incorporate new technologies. They lack the time, and, especially for newer technologies, the in-depth knowledge of the technology, required to develop high quality lessons. Technology is much more likely to be used if specific lessons, or collections of lessons, are provided in addition to the technology.

Even when collections of individual lessons are available, integrating these lessons into the existing curriculum is still an issue. As one teacher said to me after a workshop on how to use motion detectors with graphing calculators to help students understand distance-rate-time graphs, "These lessons are great. I'd like to use them in my classroom. But if I put these lessons into the curriculum, what do I take out? What do they replace?" Again, this should not be an individual teacher decision; it needs to be a departmental or district decision to maintain coherence and consistency in curricula. In addition, even more valuable than individual lessons are "replacement units" about specific content that could be incorporated into the curriculum in place of the regular instructional materials. The best option is an entire program (e.g., CME Project Algebra 1) that integrates technology throughout, and that addresses the resulting changes in curriculum value of particular topics.

Professional Development

For teachers to effectively integrate technology into their instruction requires sustained professional development and the support of school and school district leadership. While developers can provide initial professional development regarding how to use into use technological tools, actual use in the classroom involves logical and management issues in addition to knowing how to use technology itself. It is well documented that in the implementation of any innovation, teachers experience different "levels of concern" as described by the Concerns-Based Adoption Model (CBAM) – personal concerns, management concerns and concerns about effectiveness (Loucks-Horsely, 1996). Innovations involving the use of technology can exacerbate these concerns. Having the opportunity to regularly discuss technology implementation with colleagues—including management tips and students' difficulties using technology, as well as mathematical issues that arise—is essential for sustained, integrated technology use. This type of professional development support is most likely to occur when it is supported by school and or school district leadership.

IMPLEMENTATION OPPORTUNITIES

Despite the challenges and issues described above, the current environment provides several opportunities to incorporate replacement units and lessons featuring these new technologies into high school mathematics programs.

Fourth-Year Mathematics Course

As more states are increasing their graduate requirements to include four years of high school mathematics, there is growing interest in developing alternatives to the standard precalculus course as students' fourth mathematics course. At the same time, there are few existing models for such courses. Thus, there is opportunity for technology developers to partner with schools and/or school districts to develop technology-rich courses as possible fourth-year options. Such an option might be a capstone course, such as the one that Keith Jones described based on the study of significant real world and abstract mathematics problems, using dynamic geometry and CAS technologies. Or, if developing an entire course is too large an undertaking, the fourth year course is a place to try out replacement units that incorporate new technologies.

Increasing interest in "double-dose" Algebra 1 courses to address the needs of underprepared Algebra 1 students—courses in which students receive two daily periods of algebra instruction to provide time for both algebra instruction and for instruction to address gaps and misconceptions in their prior knowledge—also provides opportunities to try technology innovations. Similar to the fourth-year course situation, there are few instructional materials designed for use in such courses; generally, teachers use their school's/district's adopted algebra program for the algebra instruction, and cobble together other resources on their own to provide remedial instruction. Such courses provide an opportunity to explore how to use technology to both improve the algebra instruction and address students' misconceptions from past instruction. Such courses might also be interesting contexts in which to investigate whether the resequencing of concepts and skills has similar benefits for underprepared Algebra 1 students as for applied calculus students—increased conceptual understanding while still developing adequate computational skills (Heid, 1988).

Professional Learning Communities

Increasing numbers of schools and school districts are instituting Professional Learning Communities (PLCs)—teams of professionals who meet regularly to engage in joint work around issues of curriculum, instruction, assessment and other professional issues—to increase the effectiveness of professional practice and, as a result, increase student learning. A key issue in forming such teams is identifying the "common work" in which the teacher teams will be engaged. PLCs are most effective when the collaborative work results in specific products, e.g., identifying common learning outcomes for a particular course on a unit-by-unit basis, developing common assessments, sharing teaching strategies and analyzing their effectiveness. Implementing technology innovations could be a valuable focus of PLC work, making the growing interest in developing PLCs yet another opportunity to initiate and incorporate technology innovation into current practice.

Summary

Bridging the gap between the technology-rich curriculum and instruction described during this conference and the realities of current U.S. classrooms, especially at the high school level, poses a number of issues and challenges. An important next step in addressing this gap is to consider schools and school districts as the units of change, rather than individual teachers. Support of school and school district leadership is essential for addressing issues regarding curriculum, instruction, and professional development that are inherent in integrating the technology innovations described here into typical classrooms. Despite these challenges, the current environment offers a number of opportunities to promote implementation of these new technologies, including increasing interest in fourth-year course options other than precalculus, double-dose algebra classes to address the needs of underprepared students, and PLCs.

REFERENCES

Cuoco, A. (2008). Algebra in the age of CAS: Implications for the high school curriculum: Examples from the CME Project. This volume.

Heid, M. K. (1988). Resequencing skills and concepts in applied calculus using the computer as a tool. *Journal for Research in Mathematics Education, 19*(1), 3–25.

Heid, M. K. (2008). A perspective on the future computer algebra systems in school algebra. This volume.

Stacey, K. (2008). CAS and the future of the algebra curriculum. This volume.

Loucks-Horsley, S. (1996). Professional development for science education: A critical and immediate challenge. In R. Bybee (Ed.), *National standards & the science curriculum.* Dubuque, IA: Kendall/Hunt.

National Council of Teachers of Mathematics (1989). *Curriculum and evaluation standards for school mathematics.* Reston, VA: Author.

CHAPTER 19

TOOLS, TECHNOLOGIES, AND TRAJECTORIES

Douglas H. Clements

The conference activities, especially the presentations and discussions, were interesting and enjoyable without exception. I am honored to react to them. Of the many discussed, I restrict myself to two themes, curriculum tools, including technology, and trajectories—research trajectories and curriculum-based learning trajectories.

TOOLS AND TECHNOLOGIES

Calculators. Divergent opinions were expressed regarding the role of cultural tools for learning, from low-tech supports such as physical manipulatives to complex electronic technologies. One example was Elizabeth Warren's posing of the familiar question, "If a calculator can do it, why teach it?" Similar sentiments were heard from several during discussions (although some were more moderate). Such a position stands in stark contrast to my own review of calculator use for the National Math Advisory Panel (NMP, 2008)—at least to the Panel's re-phrasing of that review, a re-phrasing that I strongly resisted.

> A review of 11 studies that met the Panel's rigorous criteria (only one study less than 20 years old) found limited or no impact of calculators on calcula-

Future Curricular Trends in School Algebra and Geometry: Proceedings of a Conference. pages 259–266

> tion skills, problem solving, or conceptual development over periods of up to one year. This finding is limited to the effect of calculators as used in the 11 studies. However, the Panel's survey of the nation's algebra teachers indicated that the use of calculators in prior grades was one of their concerns. The Panel cautions that to the degree that calculators impede the development of automaticity, fluency in computation will be adversely affected (p. xxiv).

The "cautions" were added by Panel members, not by myself, and were accepted by the entire Panel, but not unanimously. I shall return to other limitations in this summary.

Second, note that the Panel initially agreed that the fields of educational technology and research on its effect were quite broad, and thus I should conduct a review of previous meta-analyses. This review—summarized in the full "Instructional Practices" report—was moderately but definitely positive about technology, including the use of calculators[1]. One can speculate about whether such positive results led to the many members of the NMP demanding that this review-of-reviews was inadequate, and only "rigorous" studies (those categorized as high quality by the criteria set by the Panel before I joined it) should be included. I conducted these reviews only on those categories of technology applications that showed promise in the review-of-reviews: calculators, drill-and-practice software, tutorials, and computer programming.[2]

Third, then, those in the field will recognize that these applications do not include the most recent creations. The restriction of the research reviews to less current technology resulted from two factors, (a) such applications have existed for a sufficiently long time and thus have amassed a number of studies and (b) (to me, even more problematic) recent studies have not included designs that allow causal inferences. As the previous quotation mentioned, of the calculator studies with such designs, only one was less than 20 years old.

Fourth, I recommend that those interested in calculators—or any other issue addressed by the NMP—read the entire report (embedded in the "Instructional Practices" document), including the review-of-reviews—which includes a far greater number of studies—and has a more elaborate and nuanced discussion that was less affected by the Panel's re-phrasing. Criticisms will remain, of course, as they should—but they will be better grounded and therefore more accurate and more fruitful.

"Concrete" manipulatives. The notion of *concrete,* from concrete manipulatives to pedagogical sequences such as "concrete to abstract," is embedded

[1]Positive results were found even on calculation skills, and on problem-solving competence, but, perhaps surprisingly, not on conceptual knowledge.

[2]See the full report for these results, which were generally positive, at least on certain components of mathematics achievement.

in educational theories, research, and practice, especially in curriculum development in mathematics education. Although widely-accepted notions often have a good deal of truth behind them, they can also become immune from critical reflection. In previous reviews of the research and theory regarding concrete manipulatives we concluded that while (a) using manipulatives is generally helpful, (b) manipulatives do not guarantee meaningful learning and can be useless or worse, and (c) the field needs to re-define "concrete" and its role in mathematics education (Clements, 1999; Clements & Sarama, 2009; Clements & McMillen, 1996; Sarama & Clements, 2006, 2009).

We define *Sensory-Concrete* knowledge as the level of thinking requiring *use* of sensory material to make sense of an idea. Such material often facilitates students' development of mathematical operations by serving as material supporting children's action schemes; Jim Fey offered examples of this in his presentation. This does not mean that students' understanding is only concrete; even infants make and use abstractions in thinking about number and space. Preschoolers understand, at least as "theories-in-action," principles of geometric distance and do not need to depend on concrete, perceptual experience to judge distances (Bartsch & Wellman, 1988).

Such concrete knowledge is often contrasted with (limited) abstract knowledge. Abstract knowledge can indeed be of limited usefulness: "Direct teaching of concepts is impossible and fruitless. A teacher who tries to do this usually accomplishes nothing but empty verbalism, a parrot-like repetition of words by the child, simulating a knowledge of the corresponding concepts but actually covering up a vacuum" (Vygotsky, 1934/1986, p. 150). We call such limited knowledge *abstract-only* knowledge. However, abstraction is not to be avoided, at any age. Mathematics is *about* abstraction and generalization. "Two"—as a concept—is an abstraction. Further, even infants use conceptual categories that are abstract as they classify things, including by quantity. These are enabled by innately-specified, knowledge-rich predispositions that give children a head start in constructing knowledge. These are "abstractions-in-action," not represented explicitly by the child, but used to build knowledge. When an infant says "two doggies," she is using abstraction structures of numerosity to label a concrete situation. Thus, the situation is analogical to Vygotsky's (1934/1986) formulation of spontaneous ("concrete") vs. scientific ("symbolic") concepts in that abstractions-in-action guide the development of concrete knowledge and eventually, depending largely on social mediation, become explicated as linguistic abstractions. What of this type of knowledge, a synthesis of concrete and abstract understandings?

We define Integrated-Concrete knowledge as knowledge that is *connected* in special ways. This is the root of the word concrete—"to grow together."

What gives a sidewalk concrete its strength is the combination of separate particles in an interconnected mass. What gives Integrated-Concrete thinking its strength is the combination of many separate ideas in an interconnected structure of knowledge. For students with this type of interconnected knowledge, physical objects, actions performed on them, and abstractions are all interrelated in a strong mental structure. Therefore, an idea is not simply concrete or not concrete. Depending on what kind of *relationship* you have with the knowledge, it might be Sensory-Concrete, Abstract-Only, or Integrated-Concrete.

Comparing the two levels of concrete knowledge, we see a shift in what the adjective "concrete" describes. *Sensory-Concrete* refers to knowledge that demands the support of concrete objects and children's knowledge of manipulating these objects. *Integrated-Concrete* refers to knowledge that is "concrete" at a higher level because they are connected to other knowledge, both physical knowledge that has been abstracted and thus distanced from concrete objects and abstract knowledge of a variety of types. Such knowledge consists of units that "are primarily concrete, embodied, incorporated, lived" (Varela, 1999, p. 7).

Thus, what ultimately makes Integrated-Concrete ideas—and thus also manipulatives—useful are not their immediate connections to the physical. Good uses of manipulatives are those that aid students in building, strengthening, and connecting various representations of mathematical ideas. This opens the door to consider computer manipulatives as potentially just, or more, useful for learning as physical manipulatives, a point to which I will return.

These ideas are consistent with the uses we heard about at the conference. Elizabeth Warren cited Bruner's seminal enactive ("concrete") → iconic (pictorial) → symbolic sequence. She has put that sequence to good use—such structures are useful when research-based learning trajectories (see the next section) have yet to be created. If we understand the enactive period as one of sensory-concrete knowledge, and understand that (a) symbolic knowledge—if only nascent abstractions in the form of intuitive use of (symbolic) language—is a component of that knowledge, (b) we must ensure manipulatives serve as symbols of mathematical ideas from the beginning, (c) the subgoal is to move to drawings and (other) symbols as soon as possible, and (d) the goal in integrated concrete knowledge, then this type of structure can be a good place to start. Warren's research on early algebraic reasoning reflects these subtle understandings. In lesser hands, however, the Brunerian sequence can miss the role of abstraction and symbolization in the early phases of learning and teaching.

Computer manipulatives. As noted, computers might provide representations that are just as personally meaningful to students as physical objects. Perhaps paradoxically, research indicates that computer representations may even be more manageable, "clean," flexible, and extensible than their physical counterparts. We summarize the advantages of using computer manipulatives in two broad categories: those that offer mathematical or psychological benefits to the student and teacher, and those that offer practical and pedagogical benefits.

Mathematical/psychological benefits include (a) bringing mathematical ideas and processes to conscious awareness; (b) encouraging and facilitating complete, precise, explanations; (c) supporting mental "actions on objects"; (d) changing the very nature of the manipulative; (e) symbolizing and making connections; (f) linking the concrete and the symbolic with feedback; (g) recording and replaying students' actions. *Practical /pedagogical benefits,* which help students in a practical manner or provide pedagogical opportunities for the teacher, include (a) providing another medium, one that can store and retrieve configurations; (b) providing a manageable, clean, flexible manipulative; (c) providing an extensible manipulative; and (d) recording and extending work. Such benefits, used intentionally by educators, facilitate students' construction of Integrated-Concrete knowledge.

These advantages were well illustrated by work with several environments, including those discussed by Keith Jones, Claudi Alsina, and others. Especially noteworthy were the environments shown by Nick Jackiw, Colette Laborde, and Jean-Marie Laborde, which illustrate that models of ("concrete") phenomena (in which the external visual representations are an essential component of a distributed cognition system yet to be internalized) can become models of (and symbols for) mathematics concepts, building Integrated-Concrete knowledge at a high level. That is, it may be that concepts and skills of different natures (concrete/symbolic) "co-evolve"; the simultaneous elicitation and instantiation of many levels is most successful in most circumstances.

TRAJECTORIES—RESEARCH TRAJECTORIES AND LEARNING TRAJECTORIES

The conference again illustrated the impressive knowledge and creativity of our colleagues doing curriculum development in mathematics education. I argue that such work would be better served and more influential if it were embedded within a comprehensive curriculum research framework (Clements & Sarama, 2004a). This is *not* to say that traditional strategies such as market research and research-to-practice models are preferred to the creative developmental work done, often using design experiments. It is to say that early qualitative work and even design experiments are insufficient *alone* (although more appropriate and useful than the aforementioned tra-

TABLE 1. Goals of Curriculum Research (adapted from Clements, 2007)

	Practice	Policy	Theory
Effects	• Is the curriculum effective in helping children achieve specific learning goals? Are the intended and unintended consequences positive for children? (What is the quality of the evidence?—Construct and internal validity.) • Is there credible documentation of both a priori research and research performed on the curriculum indicating the efficacy of the approach as compared to alternative approaches?	• Are the curriculum goals important? • What is the effect size for students? • What effects does it have on teachers?	• Why is the curriculum effective? • What were the theoretical bases? • What cognitive changes occurred and what processes were responsible? That is, what specific components and features (e.g., instructional procedures, materials) account for its impact and why?
Conditions	• When and where?—Under what conditions is the curriculum effective? (Do findings generalize?—External validity.)	• What are the support requirements for various contexts?	• Why do certain sets of conditions decrease or increase the curriculum's effectiveness? • How do specific strategies produce previously unattained results and why?

ditional strategies). A complete curriculum development program should address two basic issues—effect and conditions—in three domains, practice, policy, and theory, as described in Table 1.

This is why Julie Sarama and myself have suggested the use of a full curriculum research trajectory, that is, research that follows the phases of a complete *Curriculum Research Framework (CRF)* (Clements, 2007; Sarama & Clements, in press). We believe that qualitative methods are the most important for development, but also that quantitative methodologies provide experimental results, garnered under conditions distant from the developers, that are useful in and of themselves and in that they can generate political and public support. Randomized experiments are more powerful and less biased than alternative designs and can also uncover unexpected and subtle interactions not revealed by qualitative investigations (Clements & Nastasi, 1988; Russek & Weinberg, 1993).

Although iterating through, for example, repeated design experiments might lead to an effective curriculum, this would not meet all the goals listed in Table 1. The curriculum might be effective in some settings, but not others, or it might be too difficult to scale up. Moreover, we would not know *why* the curriculum is (or in ways is not) effective. Using the CRF not only documents if the design is successful in attaining achievement goals, but also traces whether that success can be attributed to the posited theory-design connections.

Maria Blanton, Diane Resek, and many others had strong and valid criticisms of the National Math Panel report, some of which concerned the Panel's rejection of qualitative research. I agree. Nevertheless, with Diane Briars, I argue that we need more evidence across the many types of evidence in our CRF, certainly not limited to, but including, randomized field trials. In my attempts to argue a point of view on many issues that I believe is shared by many at this conference, I had frustratingly few empirical sources to support my arguments. We need to know the long-range effects of our curricular creations, especially compared to a valid counterfactual on a wide range of outcomes.

Finally, our CRF has at its core the use of learning trajectories, and I argue that more intentional and explicit use of such trajectories would benefit the work of many at the conference. Learning trajectories are "descriptions of children's thinking and learning in a specific mathematical domain, and a related, conjectured route through a set of instructional tasks designed to engender those mental processes or actions hypothesized to move children through a developmental progression of levels of thinking, created with the intent of supporting children's achievement of specific goals in that mathematical domain" (Clements & Sarama, 2004b, p. 83). Such trajectories are needed to put the brilliant tools and ideas we heard about and saw in a framework. As but one example, returning to the issue with which we began, to determine whether calculator or CAS use is helpful or harmful and at what levels of learning, we need learning trajectories: a careful consideration of goals, development, and instructional tasks[3].

REFERENCES

Bartsch, K., & Wellman, H. M. (1988). Young children's conception of distance. *Developmental Psychology, 24*(4), 532-541.

Clements, D. H. (1999). 'Concrete' manipulatives, concrete ideas. *Contemporary Issues in Early Childhood, 1*(1), 45–60.

[3]Note that Jim Fey's plea is relevant here. Also, an illustration of one learning trajectory from our work is provided in the pdf file of the conference presentation at http://mathcurriculumcenter.org/conferences/CSMC/presentations.php .

Clements, D. H. (2007). Curriculum research: Toward a framework for 'research-based curricula'. *Journal for Research in Mathematics Education, 38*, 35–70.

Clements, D. H., & Sarama, J. (2009). *Learning and teaching early math: The learning trajectories approach.* New York: Taylor & Francis.

Clements, D. H., & McMillen, S. (1996). Rethinking "concrete" manipulatives. *Teaching Children Mathematics, 2*(5), 270–279.

Clements, D. H., & Nastasi, B. K. (1988). Social and cognitive interactions in educational computer environments. *American Educational Research Journal, 25*, 87–106.

Clements, D. H., & Sarama, J. (2004a). Hypothetical learning trajectories. *Mathematical Thinking and Learning, 6*(2), 163–184.

Clements, D. H., & Sarama, J. (2004b). Learning trajectories in mathematics education. *Mathematical Thinking and Learning, 6*(2), 8–89.

NMP. (2008). *Foundations for Success: The Final Report of the National Mathematics Advisory Panel.* Washington D.C.: U.S. Department of Education, Office of Planning, Evaluation and Policy Development.

Russek, B. E., & Weinberg, S. L. (1993). Mixed methods in a study of implementation of technology-based materials in the elementary classroom. *Evaluation and Program Planning, 16*, 131–142.

Sarama, J., & Clements, D. H. (2006). Mathematics, young students, and computers: Software, teaching strategies and professional development. *The Mathematics Educator, 9*(2), 112–134.

Sarama, J., & Clements, D. H. (2008). Linking research and software development. In M. K. Heid & G. Blume (Eds.), *Research on technology and the teaching and learning of mathematics: Volume 2: Cases and perspectives.* Charlotte, NC: Information Age.

Sarama, J., & Clements, D. H. (2009). *Early childhood mathematics education research: Learning trajectories for young children.* New York: Taylor & Francis.

Varela, F. J. (1999). *Ethical know-how: Action, wisdom, and cognition.* Stanford, CA: Stanford University Press.

Vygotsky, L. S. (1934/1986). *Thought and language.* Cambridge, MA: MIT Press.

CHAPTER 20

FUTURE TRENDS IN SCHOOL ALGEBRA AND GEOMETRY

Reflections on the Vision of Experts

James Fey

The strongest impression created by the presentations and discussions at the CSMC conference on "Future Trends in School Algebra and Geometry" is that existing and easily forecast computing technology has the greatest potential to reshape those school subjects. The plenary presentations demonstrated some remarkable tools for transforming what and how we teach in the core strands of the K–12 curriculum. Speakers demonstrated powerful computer algebra systems (CAS) now available in handheld computers and software that extend dynamic geometry possibilities to three dimensions. Many of the same speakers provided intellectual frameworks for thinking about ways that curriculum, teaching, and assessment in school mathematics could respond to the new technology-rich environment for doing, learning, and teaching mathematics.

As someone who is involved in work on the challenges of designing and implementing curriculum materials and policies, I approached the presentations with an optimistic but pragmatic perspective. In addition to my many positive impressions, I found several issues missing from or unresolved by the presentations and subsequent discussions. The comments that follow address those concerns.

Future Curricular Trends in School Algebra and Geometry: Proceedings of a Conference. pages 267–271

CURRICULUM IMPACT

The fundamental implication of the impressive software demonstrations is that the new tools for doing, learning, and teaching mathematics are making sophisticated mathematical topics and thinking accessible for more and more students. The speakers often illustrated that potential with what certainly seemed to be "high-end" examples of newly-accessible problems and work by very capable students. What I heard too little about was what this general trend implies for specific change in the algebra and geometry curriculum strands of school mathematics. In particular, I did not hear much discussion about how mathematics curriculum and teaching might be tuned to the diversity of interests, aptitudes, and achievement among the K–12 student population. For example, I personally believe that ready access to CAS software means that a large fraction of the high school student population simply does not need to master any very complex personal skill with symbol manipulation—but we really did not tackle that conjecture or make at least some tentative, testable propositions about who needs to learn what in the new environment for algebra. Given the recent trend in educational policy to require more and more of a traditional college preparatory mathematics curriculum for all high school students (e.g. Algebra II and beyond as a graduation requirement), it seems urgent to ask whether the content of those advanced requirements should be defined by traditional syllabi or whether the existing and easily forecast computational tools recommend something different.

ASSESSMENT IMPACT

We heard some very interesting ideas about the ways that technology enables development of more sophisticated algebraic and geometric thinking. But unless we find practical ways to assess achievement of those higher-order thinking skills, they will not become prominent goals of the activity in typical mathematics classrooms. Furthermore, we need to continue development of effective strategies for assessment of mathematical understanding and skill in the presence of computing tools. The range of available and "always online" technologies has changed the ways that everyone doing technical work can access information and do mathematical work. We have seen some accommodation to that condition in tests that include calculator-active items. But the computing environment is continually changing, so regular attention to the issue is required.

Failing progress on either of the assessment issues, the most promising ideas for using technology to transform the content and teaching of algebra and geometry will have a hard time penetrating the test-driven culture of school mathematics. As evidence in support of that claim, consider the fate of recommendations to emphasize estimation and mental arithmetic as important skills in a world of ubiquitous calculating tools. As plausible as that recommendation might be, it has turned out to be very difficult to incorporate in conventional testing instruments. There is actually less attention today to estimation and mental arithmetic than one heard nearly two decades ago.

VIRTUAL ENVIRONMENTS

As we took in the fascinating demonstrations of dynamic 3-dimensional geometry tools and the descriptions of results from use of simulations to teach algebraic concepts, it became clear to me that we need to know much more about how students understand the virtual realities presented by those environments. Interpreting a 2-dimensional representation of a 3-dimensional object (and its transformations from various viewer perspectives) is a nontrivial skill. Designing effective teaching sequences for such learning will require careful study of ways to mix hands-on and computer-based experiences.

There is a common general assumption among mathematics teachers that hands-on experience is an essential prerequisite for knowledgeable technology-based mathematical work. That may be true. But it is also plausible that experience in specially-designed computing environments can be effective in places that hands-on work is awkward, time-consuming, or of limited generality.

ALGEBRAIC THINKING

In the same sense that one has to have geometric understanding and insight to interpret and utilize the visualization aids provided by dynamic geometry tools, several speakers suggested that intelligent use of computer algebra systems requires fundamental understanding of algebraic ideas and techniques. Just what that kind of essential personal understanding might be was not specified in the talks. A naïve and seductive interpretation of the issue would be that students need to acquire personal symbol manipulation skills before they are allowed to use the powerful tools. That sort of "do it the hard way before the easy way" principle is a strongly held belief among many mathematics teachers. However, research and development projects over the past 25 years have provided some very promising prospects. They suggest that access to technology enables early access to conceptual thinking and problem solving and even that the technology itself aids in learning complex skills.

REAL MATHEMATICS AND SCHOOL MATHEMATICS

One final point that was raised in several of the plenary and small group sessions of the conference was the claim that school mathematics is different in fundamental ways from mathematics as a scientific discipline and that this difference has significant implications for teaching. In the sense that school mathematics deals almost exclusively with mathematical ideas and techniques that have been well known and understood for many years (if not centuries), that claim is certainly true. In the sense that the aim of school mathematics is essentially transmission of known material rather than creation of new results, the claim of difference is also certainly true. However, it seems to me that we want students to develop the dispositions and habits of mind that are effective in discovery and application of mathematics by mathematicians. While we do plan learning experiences for students that facilitate their development of understanding and skill, I think most of us believe that effective instruction is that which leaves a lot of the work (and pleasure of success) to the students. Thus I would suggest caution about emphasis on the differences of school and disciplinary mathematics and instead encourage a posture that aims to make experience in school mathematics an experience of doing mathematical work.

CONCLUSION

Taken together it seems to me that the conference sessions displaying future-oriented perspectives on school algebra and geometry provided a stimulating and encouraging guide to new developments. Of course, they also implied a variety of important research and development projects needed to explore the emerging possibilities and find the best in each.

REFERENCES

Blume, G. W. & Heid, M. K. (Eds.). (2008). *Research on technology and the teaching and learning of mathematics: Vol 2: Cases and perspectives.* Charlotte, NC: Information Age.

Fey, J. T., Cuoco, A., Kieran, C., McMullin, L., and Zbiek, R. M. (Eds.). (2003) *Computer algebra systems in secondary school mathematics education.* Reston, VA: NCTM.

Guin, D., Ruthven, K., & Trouche, L. (Eds.) (2005). *The didactical challenge of symbolic calculators: Turning a computational device into a mathematical instrument.* New York: Springer.

Heid, M. K. and Blume, G. W. (Eds.) (2008). *Research on technology and the teaching and learning of mathematics: Vol 1: Research syntheses.* Charlotte, NC: Information Age.

CHAPTER 21

THOUGHTS FROM A CLASSROOM TEACHER

Jim Mamer

I was asked to attend the Second International CSMC Conference and present at the "practitioners" panel near the conclusion of the conference. I imagine I was chosen for this task because I have taught middle school mathematics for the past 20 years and, more importantly, because of my strong belief in teaching with a National Science Foundation (NSF)-funded middle school curriculum, *The Connected Mathematics Project* (CMP). My belief in how well kids can learn mathematics and how teachers can teach it changed forever in 1997 as I became connected with this curriculum. Over the past ten years, I have attended numerous professional development opportunities regarding the mathematics in this curriculum and how to best help students and teachers learn and deliver middle school mathematics. My classroom has been open to any mathematics educators who wish to visit and hundreds of visitors have attended. Most folks who visit believe my students are learning mathematics at a level intended by the authors of CMP and this success had led to numerous professional development opportunities including my attendance at this conference.

I took numerous things from this conference. First and foremost, it is exciting to see some things that are being attempted in the areas of geometry, algebra, and with the use of technology around the world. My curiosity has been awakened and I will continue to search for ideas that can be carefully

Future Curricular Trends in School Algebra and Geometry: Proceedings of a Conference. pages 273–276

implemented with my students and the educators I train. The sessions were very informative and at times very entertaining and thought provoking. One area that concerned me was the lack of materials that we as educators could take away from the conference and try to implement in our classrooms. I realize that this was not the purpose of this conference, but it is the direction I would like to move in this paper and it is the direction I took in my presentation at the practitioner's panel session. *Ideas* about what can be done in regards to teaching algebra, geometry, and even using technology are all well and good. But, it is my belief that to truly impact students in a controlled and measurable fashion, teachers need to be trained with materials that have been carefully designed to create in-depth understanding.

I think I can best describe what I mean through the story of a former student of mine, Rachel, who worked through a unit on area and perimeter while she was in sixth grade.[1] The development of her thinking about area progressed and her willingness to make conjectures about how to find the areas of new shapes improved because the tasks she was completing during this three-week unit were connected, developmentally appropriate, and challenging. Rachel's thinking is the type of thinking all teachers wish for in their students, although many of us don't understand how to set up in-depth lessons that connect day after day, week after week, unit after unit, to encourage this thinking.

Rachel came to class having previously studied area and perimeter. During our unit, Rachel completed activities over a two-week period that began with seeing that perimeter is the distance around things and area is the space in things. Our activities involved the ideas of kids developing bumper car floor plans. The area of these floor plans were the square tiles placed in the designs, whereas the perimeter was the rails needed to surround the floor plan. Kids made various designs and counted the area and perimeter of their designs. Teacher emphasis was placed on what units were being counted for area and for perimeter. The students were then given various designs and asked to find the area and perimeter of each. Next, the students were given different situations in which they either knew the "fixed" area of a rectangular design or a "fixed" perimeter of a rectangular design. In both of these situations, the students were asked to find all the rectangles that met these criteria. Through these activities, students learned how to find the area and perimeter of rectangles very efficiently. In fact, the students became quite confident and actually stated formulas for finding both with little help from me. We then gave the students the chance to find the areas and perimeters of triangles and then parallelograms. Kids were given grids which had several of these shapes on them and as they counted whole

[1]If possible, the reader might wish to refer to my powerpoint presentation at www.mathematicscurriculumcenter.org.

and partial squares inside each shape and counted lengths around them, they began to see relationships between each shape's length and width and the shape's areas and perimeters. Once again, the activities pushed the kids to reason, conjecture, and believe in the mathematics they were learning.

Having a strong understanding of what area and perimeter are and also having developed formulas that the kids were quite secure with led the students to their next shape, the circle. I'm going to concentrate on the area of circles because the learning Rachel demonstrated here is the point I'm trying to make. I'd like to see this learning with all children in all fields of mathematics. Rachel was given a circle that was located on an overlay of grid paper. She and all of her classmates were told that they needed to find an estimate for the area of the circle. As I walked around the room I noticed students coming up with many ways to count the squares they could see within the circle. She told me that she knew we would eventually be finding the area formula for this shape, just as we had done with other shapes in this unit, so she said she was going to find the formula today during this activity. I remember nodding and telling Rachel good luck. (I knew she would not be able to invent the formula we would later discover involving the number of radius squares that fit inside any circle.) At the conclusion of this estimation lesson, Rachel burst with joy and told me she had discovered the formula to find the area of any circle. I was quite confident she hadn't but I asked her to state what she had discovered. She said to find the area of any circle all that needs to be done is to "square" the diameter and then take seventy-five percent of that answer. I was caught off guard by Rachel's conjecture and I stalled for time by asking her to try it on several circles at home that evening.... In the meantime I thought about her conjecture and after much time realized that her conjecture of seventy-five percent of the diameter square was extremely close to the *pi*/fourths of the diameter square which would be the actual formula.

Rachel's willingness to make the conjecture came about because of the clever manner in which Rachel had been exposed to learning throughout this unit and actually this entire school year with the NSF curriculum, the *Connected Mathematics Project.* My ability to reason about her conjecture occurred because of the wonderful professional development opportunities I've been given surrounding this curriculum and the knowledge I've gained implementing it during the past ten years. This type of meaningful learning has taken place with the majority of my students since I became exposed to this challenging new way to teach mathematics. During my first ten years of teaching, I felt strong about my ability to help children learn mathematics. However, after spending three summers receiving intense professional development surrounding the *Connected Mathematics Project* and then trying to implement the curriculum as the authors intended, I began to realize what I didn't know about teaching children mathematics for my first ten years of teaching.... Also, during the past ten I have spent a great deal of my summers and school year time trying to educate any teachers who are willing to look at this type of learning. I have found it takes vast amounts of time and energy to educate current mathematics teachers on the value and effectiveness of using a program like the one I've mentioned here. But, the effort is worth it for the benefit of students across the country.

So, attending this conference was inspirational for me but also caused me concern in some ways. I walked away excited about some ideas that were shared but concerned about how these ideas might effectively, authentically, and consistently be implemented in classrooms. Ideas and philosophies are obviously very important to great teaching. However, a huge variable is placed in the hands of teachers as we try to take ideas away from workshops and use them with students. A strong beginning step to limiting this variable is to give teachers very specific, well thought out activities to implement with students. That is what I have received with the *Connected Mathematics Project*. I believe it is very difficult and nearly impossible to expect individual teachers from our country to develop their own coherent activities that meet our state's standards and also model current best practices from around the world. I do not mean to sell teachers short, I just know from my own experiences with this one curriculum and the professional development surrounding it, we all have a great deal to learn about how kids can learn mathematics.

After telling Rachel's story to the CSMC audience I closed with a sincere challenge to the group of educators in attendance. My challenge was for these great minds to take their great ideas from around the world and continue to refine them and put them into units that can be packaged for teachers to implement. Share with us how they would spend three to four weeks developing a concept they proposed at this conference. What specific activities would they use and what learning would they expect to see from students? Give us specific questions they would ask students to challenge all, to help those who struggle, and to check for adequate understanding. These are the things NSF provided when they funded curriculum like the *Connected Mathematics Project* and these are the things many folks like me now want to continue receiving. There is so much to be gained from continuing to push educators with these types of curriculum but we need leaders to continue providing us with the "latest, greatest" units they believe we should be trying to educate kids. I am so proud of the learning I provide for the "Rachel's" of my classrooms now because of the fortunate training I received. During my next ten years of teaching I hope to receive even better units from mathematics leaders around the world to reach my children.

CHAPTER 22

RESTORING AND BALANCING[1]

William McCallum

In 1973 Morris Kline (Kline, 1973) criticized the state of school algebra as he saw it:

> They are taught many dozens of such processes: factoring, solving equations in one and two unknowns, the uses of exponents, addition, subtraction, multiplication and division of polynomials, and operations with negative numbers and radicals such as $\sqrt{3}$. In each case they are asked to imitate what the teacher and the text show them how to do. Hence ...the students are faced with a bewildering variety of processes which they repeat by rote in order to master them. The learning is almost always sheer memorization.
>
> It is also true that the various processes are disconnected, at least as usually presented. They rarely have much to do with each other. While all these processes do contribute to the goal of enabling the student to perform algebraic operations in advanced mathematics, ...as far as the students can see the topics are unrelated. They are like pages torn from a hundred different books, no one of which conveys the life, meaning and spirit of mathematics. This presentation of algebra begins nowhere and ends nowhere.

[1]The title of my talk comes from the treatise that gave us the word algebra, Al-Khwarizmi's *Hisab al-jabr w'al-muqabala* (*Book of Restoring and Balancing*), c. 825 CE.

Future Curricular Trends in School Algebra and Geometry: Proceedings of a Conference. pages 277–284

Already at the time of this description of the traditional curriculum reform efforts had been under way, aimed at making algebra seem less meaningless to students by introducing the concept of a function. More recent curriculum development efforts have also emphasized functions as a central organizing principle: modeling linear and exponential growth, for example, or introducing quadratic equations in the context of motion of a falling object.

There are some good aspects to this approach. First, the use of realistic contexts can help motivate students. Second, it can help students grasp abstract concepts and make them real. Third, functions are indeed an important concept that students have difficulty understanding. The traditional curriculum often left students unable to answer the simplest conceptual questions about functions.

However, a focus on functions and their graphs can veil the ideas at the heart of algebra: symbols, expressions, and equations. Sometimes it seems we are doomed to either fetishize or push aside these beautiful ideas. My purpose in this note is to present some examples which convey "the life, meaning, and spirit" of algebra—examples which both restore algebra from its all-too-common status as a "bewildering variety of processes" to its rightful position in the realm of mathematical ideas, and to illustrate the balance between procedural fluency and conceptual understanding needed to bring it alive.

READING EXPRESSIONS

Students who go beyond algebra might be expected to make the following sorts of observations about expressions in subsequent courses, or at least to be able to see why the observations are true when their instructor makes them:

- $P\left(1+\frac{r}{12}\right)^{12n}$ is linear in P (finance)
- $\frac{n(n+1)(2n+1)}{6}$ is a cubic polynomial with leading coefficient $\frac{1}{3}$ (calculus)
- $L_0\sqrt{1-\left(\frac{v}{c}\right)^2}$ vanishes when $v=c$ (physics)
- $\frac{\sigma}{\sqrt{n}}$ halves when n is multiplied by 4 (statistics).

What combination of skills and concepts is needed to make these observations?

The first observation requires the ability to view everything after the P as a single blob which does not depend on P. This seems to be a conceptual ability. Can that ability be acquired without hands-on experience with algebraic manipulation? How would its acquisition be affected if symbolic manipulation programs replaced paper-and-pencil drill in algebra skills?

The second observation can be verified by an expansion. If, however, the observation is made in passing by an instructor eager to get to the punchline about the integral of $f(x) = x^2$, then how is the student to see the fact directly? I imagine a sort of fast-forward mental calculation, ignoring the irrelevant points. Rather than trying to sort this ability into procedural skill or conceptual understanding, one is tempted to describe it as an inseparable blend of the two (conceptual skill? procedural understanding?).

Similarly, one might see the third observation by setting the expression equal to zero and solving for v, and one might see the fourth observation by substituting $4n$ for n and doing some algebraic manipulation. But in each case there is no time for that as the instructor rushes on, and in each case there seems to be, hovering in the background, a higher level of conceptual understanding that enables one to see the fact directly. I see this higher level as analogous to the higher levels of reading comprehension that one expects students to have acquired by the time they leave high school.

THE JOY OF SYMBOL MANIPULATION

One of the most profound ideas in algebra is the idea of an equation: a statement whose truth or falsity is contingent on the values assigned to the variables. It is the contingency of equations that enables them to make "unknowns" known: the idea that they can be evaluated to true or false statements, and thereby point to the value of an unknown without directly assigning that value, is an important and difficult idea. Like many mathematicians, I enjoyed solving puzzles as a child: I well remember the intellectual excitement of first realizing that most of them became mechanical when translated into algebraic notation.

As an example of the beautiful machinery of algebra, consider the following proof of Ptolemy's theorem, which states that in a cyclic quadrilateral (one that can be inscribed in a circle) the product of the diagonals is the sum of the products of opposite sides. In algebraic notation, referring to Figure 1, this statement becomes the beautifully symmetric equation

$$xy = ac + bd\,.$$

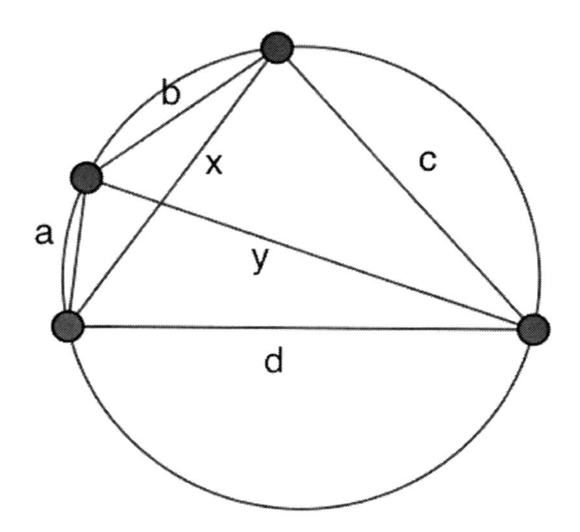

FIGURE 1. A cyclic quadrilateral

The many ways of writing this equation coming from the commutativity of addition and multiplication play an important role in the proof to come. If θ is the angle between sides a and b, then applying the law of cosines on either side of the diagonal x gives two equations:

$$x^2 = a^2 + b^2 - 2ab\cos\theta$$

$$x^2 = c^2 + d^2 - 2cd\cos\theta$$

(Here we have used the fact that opposite angles in a cyclic quadrilateral are supplementary.) Eliminating cos θ gives, after simplification,

$$x^2 = \frac{b^2cd + a^2cd + abc^2 + abd^2}{ab + cd}$$

The numerator cries out for factorization, but momentarily resists, until we rewrite it as $(bc)(bd) + (ac)(ad) + (ac)(bc) + (ad)(bd)$ and factor by grouping:

$$x^2 = \frac{(ac + bd)(ad + bc)}{ab + cd} \tag{1}$$

We could repeat the calculation for y. But it is more fun to look at the diagram and shift our perspective clockwise, so that y takes the role of x, b that of a, c that of b, d that of c, and a that of d. Making these substitutions in (1), we get

$$y^2 = \frac{(bd+ca)(ba+cd)}{bc+da} = \frac{(ac+bd)(ab+cd)}{ad+bc}$$

When we multiply the expressions for and we get some cancellation, leaving

$$x^2y^2 = (ac+bd)^2$$

Taking square roots gives us Ptolemy's theorem.

CONNECTIONS BETWEEN ALGEBRA AND GEOMETRY

My next and final example is inspired by a beautiful article in *Mathematics Teacher* (Rosenberg, Spillane, & Wulf, 2008). Working with a group of teachers, the authors of that paper were considering whether two triangles with the same area and same perimeter are necessarily congruent. The answer is no, but counterexamples are not immediately apparent, and it is particularly interesting to find pairs of triangles with rational sides which have the same area and perimeter. One such pair is the triangle with side lengths 3, 4, and 5, and the triangle with side lengths:

$$\frac{41}{51}, \frac{101}{21}, \text{ and } \frac{156}{35}.$$

Figuring out how to find such examples is a journey that leads directly from a simple question in high school mathematics to current research in number theory. Here is a brief sketch of the beginning of that journey.

We can attack the question by considering triangles circumscribed around a circle of fixed radius r, as in Figure 2. Let the triangle have area A and semiperimeter s. Then it is a pleasant exercise in geometry to see that

$$A = rs \tag{2}$$

and

$$s = r\left(\tan\frac{a}{2} + \tan\frac{b}{2} + \tan\frac{c}{2}\right) \tag{3}$$

where a, b, and c are the angles at the center of the circle formed by the radii perpendicular to the sides of the triangle.

Combining equations (2) and (3), we get

$$\frac{s^2}{A} = \frac{s}{r} = \tan\frac{a}{2} + \tan\frac{b}{2} + \tan\frac{c}{2} \tag{4}$$

Let:

$$x = \tan\left(\frac{a}{2}\right), \; y = \tan\left(\frac{b}{2}\right).$$

Note that the triangle is determined up to similarity by the pair (x, y). Since $a + b + c = 2\pi$, we have

$$\frac{c}{2} = \pi - \frac{a}{2} - \frac{b}{2}$$

so

$$\tan\left(\frac{c}{2}\right) = \tan\left(\pi - \frac{a}{2} - \frac{b}{2}\right) = \tan\left(\frac{a}{2} + \frac{b}{2}\right) = -\frac{x+y}{1-xy}$$

Referring back to equation (4) we see that, for fixed k, the equation

$$x + y - \frac{x+y}{1-xy} = k \tag{5}$$

parameterizes pairs (x, y) from which we can reconstruct triangles with

$$k = \frac{s^2}{A}.$$

Note that

$$\frac{s^2}{A}$$

is invariant under similarity, so scaling the triangle to a fixed area (say, $A = 1$) would then give us different triangles with the same area and same perimeter.

We rewrite the equation (5) as

$$kxy - x^2y - xy^2 = k \tag{6}$$

This defines a cubic curve in the xy-plane, known as an elliptic curve. Figure 3 shows the curve for the case $k = 6$. Not every point on this curve corresponds to a triangle, since there are restrictions on the angles a, b, and c, which restrict x and y to a region containing the loop-like component of the curve in the first quadrant, shown close-up on the right of Figure 3.

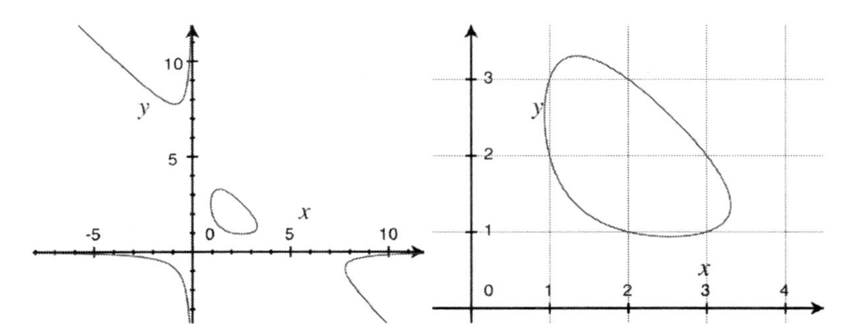

Figure 3. Curve parameterizing triangles with equal area and perimeter, k = 6.

Points on this curve with rational coordinates correspond to triangles with rational sides. For example, the (3, 4, 5) triangle is represented by six points on the curve with $k = 6$ (corresponding to the 3 choices of base and two choices of orientation). These points have coordinates (1, 2), (2, 1), (1, 3), (3, 1), (2, 3), and (3, 2). We want to find some other points with rational coordinates in order to get another triangle with the same area and perimeter as the (3, 4, 5) triangle. We can do this by the secant method: a line through two rational points intersects the curve in a third rational point. By playing around with secant method for these points you can find the point corresponding to the triangle

$$\left(\frac{41}{15}, \frac{101}{21}, \frac{156}{35}\right).$$

(In the process, you will find points on other components of the curve.) Triangles with integer sides and integer area are called Heron triangles. Obviously any triangle with rational sides and rational area can be scaled up to a Heron triangle.

Elliptic curves are the subject of current research in number theory. In particular, the curve (6) has been studied in slightly different form by van Luijk (2004), who finds infinitely many families, each of which contains infinitely many triangles with rational sides and rational area, all with the same perimeter and area.

CAN WE MAKE ALGEBRA DYNAMIC?

The previous section gives an example of one way of answering this question. Dynamic geometry software provides a rich environment for investigating geometric questions that are closely linked to algebraic questions.

Is there a way of bringing algebraic manipulations alive directly? The traditional prescription is extensive practice. But as the information environment becomes saturated with computational power, this prescription will seem more and more bizarre to students, and we need to explore others.

Is it possible to design a dynamic algebra program that allows students to read and work with expressions and equation without withdrawing cognitive access to the manipulations? To what extent can computer algebra systems replace paper and pencil calculation, while still allowing students to acquire the symbolic understanding exemplified in this paper? In "Thinking Out of the Box" (2003), I propose a methodology for approaching this question. Here I simply point out that the existence of computer algebra systems, whether you want to embrace them or ban them, increases the need to focus on the core ideas of algebra: symbols, expressions, equations, and functions.

REFERENCES

Kline, M. (1973). *Why Johnny can't add: The failure of the new mathematics.* New York: St. Martin's Press.

McCallum, W. (2003). Thinking out of the box. In J. T. Fey, A. Cuoco, C. Kieran, L. McMullin, & R. Zbiek (Eds.), *Computer Algebra Systems in Secondary School Mathematics Education* (pp. 73–86). Reston, VA: NCTM.

Rosenberg, S., Spillane, M., & Wulf, D. B. (2008). Delving deeper: Heron triangles and moduli spaces. *Mathematics Teacher, 101*(9), 656.

van Luijk, R. (2004). *An elliptic K3 surface associated to Heron triangles.* arXiv:math/0411606v1 [math.AG].

PART VI

REFLECTIONS BY CONFERENCE ATTENDEES

CHAPTER 23

INSIGHTS INTO DYNAMIC MATHEMATICAL LEARNING ENVIRONMENTS

Sarah J. Hicks, Melissa D. McNaught, and J. Matt Switzer

At the conference, the participants, including the authors of this paper, discussed important topics such as how exploration, proof, theorems, applications, and technology are incorporated into mathematics curriculum and mathematics students' learning environments. We noted one way to begin to comprehend these topics is to consider how to design learning environments using technology in order to reveal, support, and enhance students' mathematical literacy. Hence, we have chosen to organize our discussions of the conference and areas for future research around the Bransford, Brown, & Cocking (2000) model (figure 1), which captures perspectives on learning environments within four domains: community-centered, knowledge-centered, assessment-centered, and learner-centered environments (Bransford et al., 2000). Consequently, in this paper, we highlight some of the presenters' positions taken with respect to the four domains and conclude by identifying suggestions for future research and scholarly discussions.

Future Curricular Trends in School Algebra and Geometry: Proceedings of a Conference. pages 287–294

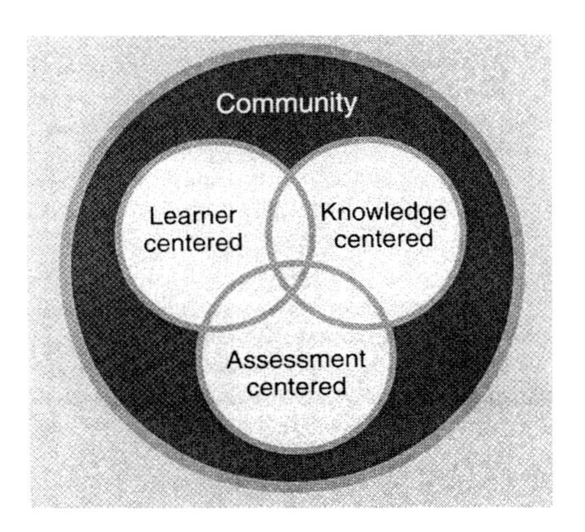

FIGURE 1. Perspectives on learning environments (Bransford et al., 2000).

ALGEBRA AND GEOMETRY IN A COMMUNITY-CENTERED ENVIRONMENT

In community-centered environments, there exist norms for people learning from one another and continually attempting to establish an environment in which it is acceptable for students (and teachers) to make mistakes during the process (Bransford et al, 2000). As participants in this community, students and teachers use their language, signs, and cultural tools to communicate and negotiate meaning with one another. During this conference, Elizabeth Warren and Al Cuoco showcased two newly developed curriculum series which have critically incorporated language, mathematical signs, and tools into the curriculum in order to help establish mathematically meaningful, community-centered learning environments in elementary and secondary classrooms, respectively.

In Australia, the Early Algebraic Thinking Project (EATP) has recently been developed and researched in order to support algebra learning at earlier school levels. One feature of this curriculum is the use of models (i.e., internal and external representations; mental, iconic, concrete, graphic, movement, and symbolic models), which provide platforms for challenging student thinking. By working between the internal and external representations, students abstract the essence/structure of mathematics work (Warren, at this conference). A large part of this is the use of language that is essential for the development of algebraic thinking as students move between representations and develop signs for dealing with abstraction. Warren claims, "Algebra is not about guessing but is about using signs in a distinctive way" (at this conference). Hence, elementary mathematics education should not simply be about basic number facts. Rather, teachers should create, understand, and/or use models (e.g., the balance model), to help students learn from one another and improve their mathematical understanding. Furthermore, technology can be used to help even young students think about mathematical signs in distinctive ways, and is changing the norms of mathematics learning in elementary as well secondary education learning environments. The use of technology would allow us to deemphasize calculation in order to focus on mathematical structure. As Warren suggests, this is important because the power of mathematics at the earlier school levels lies within the interlink of algebraic and arithmetic thinking, where arithmetic thinking focuses on products and algebraic thinking focuses on processes (C. Laborde, at this conference).

In the United States, the National Science Foundation (NSF) has recently supported the development of a high school mathematics curriculum that incorporates technology as an integral tool to advanced mathematical learning. The Center for Mathematics Education (CME) Project for high school level mathematics, authored by Al Cuoco (2009) and published by Pearson, engages students (and teachers) in a mathematical style of work and mathematical *habits of mind.* In other words, the curriculum aims to prepare students to be comfortable with ill-posed and fuzzy problems, to see the benefit of systematizing and abstraction, and to look for and develop new ways of describing situations (Cuoco, Goldenberg, and Mark, 1996). By incorporating TI-Nspire technologies into the curriculum, students reason about calculations, seek structural similarity, and reason about operations with technology as they study geometry and algebra. More specifically, the computer algebra system (CAS) capability of TI-Nspire provides a platform for experimenting, reduces computational overhead, and uses polynomials as modeling tools in the CME texts (Cuoco, at this conference). With the technology, students and teachers get immediate feedback as they create, manipulate, and communicate mathematical representations to one another or within their individual work with the technology. The new high school algebra and geometry curriculum encourages mathematical experimentation with the use of hand-held or computer technologies in community-centered learning environments. The integration of new technologies, signs, tools, and languages of CME promotes norms for developing students' mathematical habits of mind, which affect students' relationships and decision-making beyond the walls of the classroom and school.

ALGEBRA AND GEOMETRY IN A KNOWLEDGE-CENTERED ENVIRONMENT

While new community-centered curriculum development and research work is underway in the United States and Australia, knowledge-centered technology development and research work is underway in France and the United States. Collette and Jean-Marie Laborde and Nick Jackiw have created and continue to develop and enhance *Cabri* geometry and *Geometer's Sketchpad*, respectively. They argue that using technology in mathematics classrooms, study, and work can and will advance sense-making. In other words, visualization, exploration, and communication via dynamic geometry (e.g., *Cabri* geometry and *Geometer's Sketchpad*) mediate learning in ways that lead to understanding and subsequent transfer of knowledge (Bransford et al., 2000).

Collette Laborde suggests that dynamic representations such as those accessible to students and teachers in the form of dynamic geometry programs foster an operational mode (e.g., make interpretations from the representations—intermediate to illustrating concepts and operating directly on representations) for students to work within during mathematics classes. It is within this operational mode that students gain understanding and the ability to transfer knowledge. Without providing opportunities for students to work within operational modes, they are limited to procedural work with various representations and miss out on making connections between procedures and concepts. After conducting research with 15- and 16-year old students in France and Italy as well as 4th year university students, she concludes that geometric and algebraic concepts and connections are mediated by representations and the new technologies can and do link geometric and algebraic understandings in knowledge-centered environments. Hence, these ideas reinforce how dynamic geometry programs offer an environment for establishing a mathematical relationship, make possible the problem, and require users to articulate between graphical and symbolic representations (Laborde, at this conference).

ALGEBRA AND GEOMETRY IN A LEARNER-CENTERED ENVIRONMENT

Presentations and discussions of learning trajectories and students' work with algebra and geometry technology brought forth the importance of future learner-centered environments. Learners come to the classroom with knowledge and innate abilities; they are not blank slates. In a learner-centered environment, efforts are made to identify the knowledge, skills, attitudes, and beliefs that students bring to the classroom and to use the results to help learners progress toward identified learning goals. Romulo Lins noted that currently, "What we do, the way we think about our children's thinking and our views on what we want them to learn are not quite what they do, not quite the way they think about their thinking, not quite what they believe they have to learn" (Lins, at this conference). Doug Clements argued for developing learning trajectories to describe students' thinking and hypothesize potential pathways of learner development toward learning goals in order to organize these efforts.

Regardless of the curriculum being used in classrooms, it is ultimately important for students to learn and engage in mathematics because, as Lins cautions, "The failure of our students in algebra is not the failure of those who tried and failed. It is rather, I argue, the failure of those who never tried to succeed in it" (at this conference). Technology (e.g., *Geometer's Sketchpad,* computer applets) and open-ended curriculum tasks have been developed to provide a dynamic platform for students to explore, investigate, and justify algebraic and geometric ideas. Creating an active environment, technology and open-ended tasks encourage students to find their own solutions, justify reasoning, and generalize findings. Doing mathematics shifts from executing algorithms to conjecturing and exploring in order to promote mathematical sense-making. The focus of mathematics is on the discipline as a whole rather than learning its strands of algebra and geometry individually. Classroom norms must be re-established so that the student understands expectations and can be an active participant in the learning process.

ALGEBRA AND GEOMETRY IN AN ASSESSMENT-CENTERED ENVIRONMENT

In addition to the community-, knowledge-, and learner-centered learning environments, assessment-centered environments are important not only to algebra and geometry, but to education in general. Assessment needs to go beyond what a student can produce, possibly without understanding, to revealing evidence of what students know, understand, and can do. The potential of technology to impact assessment is great indeed. Technology provides the opportunity for immediate and individualized formative feedback, as well as long-term summative results. The impact of technology in this environment should also be influenced by the impact that technology may have on learning goals, which will guide what will be assessed.

Discussions during the conference regarding assessment and technology took on two broad themes. The first theme focused on whether technology should be allowed on current assessments such as classroom, state, and national-level assessments (e.g., ACT, SAT, and NAEP). The question here focuses less on the assessment itself and more on the whether the inclusion of technology is desired, warranted or worthwhile. The second broad theme addressed the design of future assessments in light of the dynamic nature of algebra and geometry technology.

SUGGESTIONS FOR FUTURE RESEARCH

Reflection upon how algebra and geometry will change due to innovation in future community-, knowledge-, learner-, and assessment-centered environments calls for reconsideration of mathematical norms in school classrooms, the dynamic nature of algebraic and geometric content, student learning trajectories, and assessments with technology. Similarly, worthwhile avenues of future research relate to curriculum, teaching, student learning, and assessment.

CURRICULUM

With opinions as diverse as Elizabeth Warren's, "If a calculator can do it, then why teach it?" to Bernhard Kutzler's suggestion that the technology need not change the content, the need for further research into technology's impact on curriculum development is needed. The timeless question of what should be included in the curriculum cycles again as we consider how technology impacts content and how algebra and geometry are linked dynamically. Currently, curricula focused on the structure of mathematics and teaching a style of work with particular habits of mind are available (e,g., Cuoco, at this conference). The implementation of these curricula offers areas of research with regard to their impact on student learning.

TEACHING

The complex nature of teaching makes it difficult to document what makes an effective teacher. However, research has established that teachers matter (Good, Biddle & Brophy, 1975). The areas of teacher education and teacher pedagogy offer two domains of promising future research (Warren, at this conference).

Teacher Education. Teachers will need strong and flexible subject matter knowledge in order to be able to move between representations (e.g., balance models, symbolic models, geometric drawings, and algebraic graphs). Teachers will need to be able to use the technologies and understand how using the technology influences the mathematics, which involves making sense of the structure and connections within mathematics. Research is needed to develop programs that will assist teachers in developing these knowledge bases.

Teacher Pedagogy. Newly designed tasks and feedback of dynamic geometry challenge teachers (C. Laborde, at this conference), as do multiple representations (Warren, at this conference), as they try to help multiple learners at a variety of cognitive levels construct their own meanings and make sense of the mathematics individually and within the larger whole-class environment. Teachers must choose between representations to assist student abstracting and identify learning moments for students. Research efforts need to make connections between theory and practice in order to be able to support the further development of teacher pedagogy.

STUDENT LEARNING

The increasing accessibility of technology is changing the potential landscape of education. However, the introduction of technology into the classroom does not guarantee learning. On the other hand, with technology, learning environments have the possibility of becoming more interactive and meaningful. Thus, learning trajectories that include student understanding of content along with students' use of technology are lacking and thus provide a fertile area for research.

ASSESSMENT

In his closing remarks, Jim Fey noted that assessment remains an insufficiently addressed topic. The introduction of technology, in all its various forms and implementations, requires mathematics education researchers to address the question of how assessment will occur in these technology-rich environments. As mentioned earlier, this goes beyond the question of whether technology should be included in current assessments. Instead, the question of how we design assessments that meaningfully take advantage of the technology available, help guide the development of future technology, and are both formative and summative in nature needs to be researched. In addition, closely tied to the knowledge-centered environment, the question of what mathematics will need to be assessed in the near future also arises.

CLOSING

As curricular trends in school algebra and geometry change, we must consider their purposes for the study of mathematics and what mathematical goals we wish to achieve. For example, what actions should students engage in to promote mathematical understanding? Should students *explore* within mathematics? Should students work to *secure* and automate mathematics knowledge? Should students *apply* mathematical knowledge? Do we want to develop and support creative students and thinkers? The challenge for the future, then, is designing and providing mathematical learning environments that support creative and knowledgeable students (Kutzler, at this conference). Further research in student learning, curriculum, teaching, and assessment should guide the design of our ever-changing school mathematics learning environments.

REFERENCES

Bransford, J. D., Brown, A. L., & Cocking, R. R. (Eds.). (2000). *How people learn: Brain, mind, experience, and school.* Washington D.C.: National Academy Press.

Cuoco, A., Goldenberg, E. P., & Mark, J. (1996). Habits of mind: An organizing principle for mathematics curricula, *Journal of Mathematical Behavior, 15*, 375–402.

Cuoco, A., et al. (2009). *CME Project.* Newton, MA: Pearson.

Good, T. L., Biddle, B. J., & Brophy, J. E. (1975). *Teachers make a difference.* New York: Holt, Rinehart, and Winston.

CHAPTER 24

INSTRUMENTAL GENESIS AND FUTURE RESEARCH IN SCHOOL ALGEBRA AND GEOMETRY

Daniel J. Ross

INTRODUCTION

Discussion related to technology was ubiquitous at the Second International Mathematics Curriculum Conference addressing future trends in school algebra and geometry. Each researcher brought a unique perspective to the role of technology in mathematics education. The theory of instrumental genesis provides a useful framework for unifying these discussions and providing insight into their arguments. In addition, examining their discussions through the theory can reveal areas for potential future research.

INSTRUMENTAL GENESIS

Speakers at the conference suggested that computer technology is playing an ever-increasing role in school mathematics, and other researchers

Future Curricular Trends in School Algebra and Geometry: Proceedings of a Conference. pages 295–304

have also made this argument. Zbiek, Heid, and Blume (2007) noted the explosive growth of the availability technology for mathematics classrooms over the past 25 years. Clements and Battista (2000) wrote, "There is little doubt that technology will have a major impact on the teaching and learning of mathematics" (p. 761). Drijvers (2003) argued that technological tools, specifically computer algebra systems, are becoming more common in classrooms. Ferrara, Pratt, and Robutti (2006) claimed that over the last 30 years "technology has shaped the way algebra is perceived" (p. 238). They noted how technology has allowed students to engage with algebraic concepts dynamically and to see more immediate connections between representations.

A useful construct for examining the question of how students come to learn mathematics through technology is the theory of instrumental genesis, which Zbiek, Heid, and Blume (2007) noted as a promising way to better understand relationships between students and tools. The theory of instrumental genesis did not begin within the mathematics education arena, but instead emerged from French ergonomic theory. Initially, Verillon and Rabardel (1995), described a general situation in which a subject uses an instrument to access or act on an object, which they call an Instrumented Activity Situation (IAS). They indicated that the three constructs of subject, instrument, and object act on and influence each other as the subject strives to achieve his or her goal (see Figure 1). This model highlights the variety of interactions that connect the three constructs involved in the situation as well as the intermediary nature of the instrument.

The theory of instrumental genesis hinges on the concept of instrument. To Verillon and Rabardel (1995), an instrument was not simply a physical or symbolic object, but a complex psychological structure with two major components. The first is the material component, which they called the

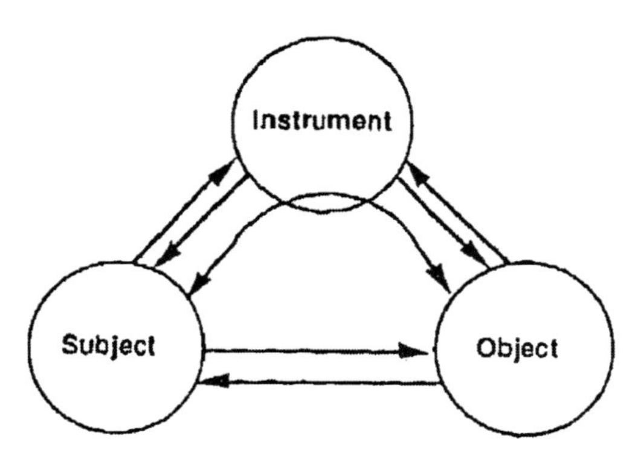

FIGURE 1. The IAS

artifact. An example of an artifact discussed frequently at the conference was dynamic geometry software. Such software generally has a visual representation that consists of symbols displayed on a computer screen. It has an external structure, which consists of the graphical user interface, keyboard, and mouse that users interact with, as well as an internal structure related to the rules governing the operation of the software.

In addition to the artifact, the other major component of an instrument is the subject's understanding, representations, cognitive operations, and motor skills used in relation to the artifact. Together these can be described as a system of utilization schemes associated with the artifact. These schemes are stable cognitive structures that allow the subject to anticipate and generalize outcomes of actions taken with the artifact. Guin and Trouche (2002) described schemes as "the invariant organization of the behaviour in a given class of situations" (p. 205). These schemes relate to the subject's understanding of what the artifact can be used for, the conditions that influence its use and outcomes, and the way the subject should interact with the artifact. These schemes are also related to the specific artifact (or class of artifacts) and the object of the IAS.

An interesting example to highlight the concept of scheme from the conference was presented by Jean-Marie Laborde. In a new version of *Cabri*, the designers have included virtual tools that look and work like their real counterparts rather than tools in earlier versions of *Cabri*. For example, the tool for making a circle has a graphical representation that resembles a compass and can be used like a compass. Previously, students with experience with a compass would have certain schemes related to how the tool could be used to help them approach geometrical problems. They would have certain expectations for how the tool could be used and what outcomes would be. However, when they used *Cabri*, they would need to develop different schemes to use the circle tool that did not look or act like a compass. Now, with the tool in *Cabri* being more closely related to its physical counterpart, students' schemes for the physical and technological tools should have greater overlap.

Schemes are important because instruments are not merely material artifacts, but are psychological constructs developed through the object-oriented action of a subject in relation to an artifact. Béguin and Rabardel (2000) pointed out that the instrument is a mediator between the subject and the object; however, they also claimed that the instrument is actually made up of the subject and the object. The artifact and the formation of the subject's schemes are bound up with the object in the IAS. With a different object, a different instrument would be formed.

Instrumental genesis is the process of subjects developing schemes as they use and transform artifacts (Verillon & Rabardel, 1995). Through this process, instruments are formed. Researchers have generally focused on two dimensions of the process, instrumentalization and instrumentation. Béguin and Rabardel (2000) noted that both dimensions relate to the subject; however, the direction in which they occur is oriented toward the artifact and the subject, respectively.

Through instrumentalization the artifact becomes enriched via the actions of the subject (Béguin & Rabardel, 2000). As the subject uses the artifact, he or she engages with the object, and the resulting instrument retains this functionality, whether for a short period or for longer times. The artifact may, but need not be, physically changed. The evolution is mainly psychological. As an example from the conference, Jean-Marie Laborde spoke about how Cabri 3D can be adjusted to create different representations of three-dimensional objects. A student may not initially know how to interpret and switch between these representations. However, as the student comes to know the software, he or she can learn to use the options available in the software to change representations, making changes to enrich the artifact.

Béguin and Rabardel (2000) and Trouche (2004) describe three hierarchical levels of instrumentalization. The lowest level involves discovery and selection of functions and relates only to a single action in a highly-specified situation in which the action occurred. The instrument is temporary. At the next level, the user personalizes the artifact and it is retained for the long term as an instrument and is useful in broader situations. At the highest level, the artifact is transformed and the changes in functionality become integral and permanent to the artifact itself through relatively permanent modification to its material structure.

The other main dimension of instrumental genesis is instrumentation. Through instrumentation, the subject develops utilization schemes. Béguin and Rabardel (2000) argued that there are two processes involved in instrumentation. In an accommodative process, schemes as active structures change as the user engages with the artifact. Verillon and Rabardel (1995) argued that it is due to the expansion of possible actions afforded by the artifact which enable changes in users' schemes. Béguin and Rabardel explained that in contrast, in an assimilative process, a scheme could be applied to multiple artifacts that share common features. For example, a student familiar with Geometer's Sketchpad would be comfortable using many features of Cabri as well. Béguin and Rabardel also noted that schemes have both an individual and a social dimension. Individuals form personal schemes as they use artifacts, and societies develop shared schemes transmitted through culture and education. Verillon and Rabardel (1995) noted the essential social nature of the process due to the interaction of users with each other and with the designer through instructions for use of the artifact.

A main tenet of the theory is that instrumentalization and instrumentation intertwine and influence each other. The features of the artifact influence the utilization schemes that a user develops. Trouche (2004) wrote, "[an] artifact *prints its mark* on the subject" (p. 290, emphasis in the original). A user's schemes also affect the way she or he adapts the artifact. As Trouche (2004) wrote, "*the user's conception of the instrument is formed through use*" (p. 295, emphasis in the original). This interaction generates complexity in the process of genesis recognized by many researchers (Artigue, 2002; Guin & Trouche, 1999, 2002; Haspekian, 2005; Hoyles, Noss, & Kent, 2004; Mariotti, 2002; Trouche, 2004). The process of students learning to use technological artifacts in mathematics classrooms is not straightforward. The theory of instrumental genesis can give researchers a frame to approach questions related to technology in the mathematics classroom. In the following section, it will be used to examine discussions from the conference related to technology.

INSTRUMENTAL GENESIS AND TECHNOLOGY THEMES FROM CONFERENCE PRESENTATIONS

No presenters directly spoke about instrumental genesis at any length; however, several themes related to use of technology to teach algebra and geometry emerged in presentations throughout the conference. The theory of instrumental genesis provides a perspective that offers insight into connections between speakers' comments and suggests approaches to the questions they raised. Three themes are examined in light of the theory.

Representations

Several speakers highlighted ways that technological tools affect the creation and understanding of representations. The theory of instrumental genesis suggests that areas of interest related to representations include artifacts' specific design aspects that govern how a representation is generated as well as how students' schemes related to representations affect their use of technology. Claudi Alsina argued that when learning about three-dimensional objects, technology only provides a representation of reality, not reality itself, suggesting that students' schemes related to use of technological tools will not match their schemes related to use of real objects. As noted above, Jean-Marie Laborde discussed representing 3D objects with 2D displays on technological tools. He emphasized that students' schemes may not be as well developed as mathematicians' and, therefore, students will need more experience with various representations. Stated in terms of the theory, they are at an earlier stage of instrumental genesis. Nicholas Jackiw argued that when working with students on a given mathematical concept, initial focus needs to be on developing fluency and comfort with one representation before introducing other representations. In terms of instrumental genesis, this would suggest that students should be encouraged to develop rich schemes with one type of representation before complicating the genesis with a multitude of representations. Jim Mamer, a classroom teacher, described some students' difficulties with understanding technological representations as opposed to physical representations. Examining his viewpoint through instrumental genesis, aspects of the technological artifacts and the representations they generate interferes with some students' development of connections to their schemes related to the physical objects being represented. Limitations of the artifact thus limit the genesis of an instrument. Finally, Doug Clements also pointed out that we need to seek to better understand the effects caused by features of technological artifacts. He brought up an example of how software simulating a puzzle always oriented representations of pieces correctly for students, whereas pieces of an actual puzzle might be in any orientation. Thus, if a student only used the technology to work with the puzzles, their schemes might not be sufficient to deal with a similar situation in real life. Examining discussion of technological representations through an instrumental genesis viewpoint highlights the importance of student schemes related to representations and how design of a given artifact shapes how students understand and use representations.

Changes in the Mathematics Curriculum

Another theme frequently addressed was how technology could allow and even encourage changes in the mathematics curriculum. Jim Fey argued that many trends recently in school algebra and geometry have been driven by technology and that we need to explicate what kind of algebraic understanding is needed to use computer algebra systems (CAS) wisely and understand what it is doing. From the viewpoint of instrumental genesis, he was arguing that the artifacts involved are changing what students learn and are expected to learn, but that we also need to attend to the role of students' schemes as they form instruments with artifacts such as CAS. Kaye Stacey cited the Curriculum Value of a Topic framework that could guide curriculum changes due to CAS. The framework included epistemic, pedagogical, and pragmatic aspects. Instrumental genesis informs each of these by encouraging questions about how the design of the CAS and students' schemes might affect each of these aspects. Al Cuoco spoke more specifically about how CAS could allow students to build computational models of algebraic objects that have no physical counterparts but can be used to illuminate algebraic structure. In terms of instrumental genesis, this implies that the CAS artifact can allow students to build schemes and develop an instrument that they could not develop with a physical artifact. Collette Laborde made a similar argument with an example related to the concept of function. She argued that dynamic geometry software could allow students to grasp a geometric notion of function that would better highlight the dependence relation between the variables. Students would be able to develop schemes related to function through use of the dynamic geometry artifact that would lead to the genesis of an instrument to better understand the dependency relationship in functions. Nicholas Jackiw made the broadest claim regarding changes in mathematics curriculum by arguing that dynamic software could reshape how we understand relationships between algebra and geometry and that such a new representational infrastructure would mean an entirely new mathematics. This claim fits with the core concept of instrumental genesis: that the characteristics of an artifact and the understandings we bring to it can generate an entirely new instrument. Thus, the dynamic nature of the software, combined with a reconception of algebra and geometry, allow for different mathematical thinking. The power of technology to enable students to access different mathematics is essentially what each of these researchers was speaking about, but at different levels of specificity.

Changes in Pedagogy

In addition to changes in curriculum, a theme related to changes in teaching due to technology emerged in researchers' presentations. Bernhard Kutzler argued that technology can make new pedagogy possible and that these changes in teaching, along with how we ask students to use technology, need to be driven by our goals for students. He pointed out that technology allows students to explore problems and situations, an aspect often missing in current teaching practices. In terms of instrumental genesis, the characteristics of an artifact create opportunities for students to develop different schemes and for different instruments to form. Thus, new pedagogies are possible. Specifically, students have instruments that enable them to explore mathematics. The Instrumented Activity Situation (IAS) and the genesis itself are object-oriented, and thus these objects shape students' use of technology. Al Cuoco gave a specific example of pedagogical change related to using CAS to teach about algebraic structures. He demonstrated how using CAS in teaching secondary mathematics could provide students with opportunities to explore situations and make mathematical deductions. In this case, as students bring their schemes to interactions with the CAS artifact, an instrument is formed that allows them to approach new problems and develop new understandings. M. Kathleen Heid also provided an example of how teaching with CAS created an opportunity for a teacher to ask a student to explore properties of rational functions. Because the student had developed an instrument with the CAS artifact, he was able to use it to help him examine the problem and monitor his own understanding. Jean-Marie Laborde also noted that the teacher's own familiarity with the technology affects their pedagogy and use of the technology in the classroom. From an instrumental genesis viewpoint, the teacher's own instrumental genesis with the technology is thus important because it impacts how they guide students' instrumental geneses. The theme emerged across discussions that the process of instrumental genesis is important because it affects how teachers can teach, how students learn, and especially students' opportunities to explore mathematical situations.

FUTURE CURRICULUM RESEARCH

Examination of themes emerging from the conference in relation to use of technology in school algebra and geometry through the perspective of instrumental genesis not only provides insight into the themes, but also suggests areas for future research. In relation to the theme of representation, future research should examine the relationship between particular features of technological artifacts, representations of specific mathematical topics, and student understandings and schemes related to the artifacts and the representations. For example, the new version of *Cabri* offers a compass tool with a representation very similar to a physical compass, as opposed to the circle tool of previous versions. What impact will this have on student thinking related to use of this tool and related to the use of an actual compass? What are differences in development of students' understanding when using the new virtual compass contrasted with those using dynamic software with the previous circle tool? Such questions can be raised about the many graphical, symbolic, numeric, and verbal representations in CAS, dynamic geometry software, graphing calculators, and other tools.

Future research should also examine how specific aspects of technological artifacts influence the mathematics of the school curriculum. For example, some might believe the capability of CAS to factor polynomial expressions reduces the need for students to learn to factor polynomial expressions in algebra. However, research needs to pursue more specific details. Not all CAS are the same. Particular aspects of each CAS affect how students use the tool and what might be changed in the curriculum. Also, few would argue that the concept of factoring is unimportant in algebra and higher mathematics. What features of a given CAS support student understanding of factoring and in what ways? Researchers need to study how changes in curriculum in conjunction with features of a given technological artifact affect student understanding.

The theory of instrumental genesis also suggests that future research needs to examine how specific technological artifacts interact with specific pedagogical techniques. Several presenters suggested that technology allows students to explore mathematical concepts. This area should be pursued in relation to particular combinations of technology and mathematical topics. For example, what aspects of *Cabri* support or hinder student exploration of rigid transformations? What features of *Cabri* guide or limit student exploration and what effect does this have on their understanding of rigid transformations? Research must also pursue questions regarding the teacher's role in use of technology. What role should the teacher take on as students and artifacts develop into instruments? What impact does the teacher's own instrumental genesis with a given artifact have on students' instrumental geneses?

CONCLUSION

There is little question that technology will play an important role in future algebra and geometry classrooms. The theory of instrumental genesis can provide a useful framework for studying relationships between students, mathematics, and technology in the classroom. The theory provides coherence and insight into discussions presented at the conference. In addition, viewing the themes that emerged at the conference through the lens of instrumental genesis reveals several areas for potential future research.

REFERENCES

Artigue, M. (2002). Learning mathematics in a CAS environment: The genesis of a reflection about instrumentation and the dialectics between technical and conceptual work. *International Journal of Computers for Mathematical Learning, 7*(3), 245–274.

Béguin, P., & Rabardel, P. (2000). Designing for instrument-mediated activity. *Scandinavian Journal of Information Systems, 12,* 173–190.

Clements, D. H., & Battista, M. T. (2000). Designing effective software. In A. E. Kelly & R. A. Lesh (Eds.), *Handbook of research design in mathematics and science education* (pp. 761–776). Mahwah, NJ: Lawrence Erlbaum.

Drijvers, P. (2003). Algebra on screen, on paper, and in the mind. In J. T. Fey, A. Cuoco, C. Kieran, L. McMullin & R. M. Zbiek (Eds.), *Computer algebra systems in secondary school mathematics education* (pp. 241–267). Reston, VA: NCTM.

Fennell, F., Faulkner, L. R., Ma, L., Schmid, W., Stotsky, S., Wu, H.-H., et al. (2008). *Report of the task group on conceptual knowledge and skills.* Retrieved May 28, 2008, from http://www.ed.gov/about/bdscomm/list/mathpanel/report/conceptual-knowledge.pdf.

Ferrara, F., Pratt, D., & Robutti, O. (2006). The role and uses of technologies for the teaching of algebra and calculus. In A. Gutiérrez & P. Boero (Eds.), *Handbook of research on the psychology of mathematics education: Past, present and future* (pp. 237–273). Rotterdam: Sense Publishers.

Guin, D., & Trouche, L. (1999). The complex process of converting tools into mathematical instruments: The case of calculators. *International Journal of Computers for Mathematical Learning, 3,* 195–227.

Guin, D., & Trouche, L. (2002). Mastering by the teacher of the instrumental genesis in CAS environments: Necessity of instrumental orchestrations. *Zentralblatt für Didaktik der Mathematik, 34*(5), 204–211.

Haspekian, M. (2005). An "Instrumental approach" to study the integration of a computer tool into mathematics teaching: The case of spreadsheets. *International Journal of Computers for Mathematical Learning, 10*(2), 109–141.

Hoyles, C., Noss, R., & Kent, P. (2004). On the integration of digital technologies into mathematics classrooms. *International Journal of Computers for Mathematical Learning, 9,* 309–326.

Mariotti, M. A. (2002). The influence of technological advances on students' mathematics learning. In L. D. English (Ed.), *Handbook of international research in mathematics education* (pp. 695–723). Mahwah, NJ: Lawrence Erlbaum.

National Council of Teachers of Mathematics. (2000). *Principles and standards for school mathematics.* Reston, VA: Author.

Stump, S. L. (2001a). Developing preservice teachers' pedagogical content knowledge of slope. *Journal of Mathematical Behavior, 20,* 207–227.

______. (2001b). High school precalculus students' understanding of slope as measure. *School Science and Mathematics, 101*(2), 81–89.

Trouche, L. (2004). Managing the complexity of human/machine interactions in computerized learning environments: Guiding students' command process through instrumental orchestrations. *International Journal of Computers for Mathematical Learning, 9,* 281–307.

Verillon, P., & Rabardel, P. (1995). Cognition and artifacts: A contribution to the study of thought in relation to instrumented activity. *European Journal of Psychology of Education, 10*(1), 77–101.

Zbiek, R. M., Heid, M. K., & Blume, G. W. (2007). Research on technology in mathematics education: The perspective of constructs. In J. Frank K. Lester (Ed.), *Second handbook of research on mathematics teaching and learning* (Vol. 2, pp. 1169–1207). Charlotte, NC: Information Age.

CHAPTER 25

CLOSING REMARKS

Reflections from a Retiring Mathematics Curriculum Developer

Zalman Usiskin

At this point in every conference it is appropriate to thank those individuals without whose work the conference would not have taken place. There are four groups of individuals:

- the planning committee—Sarah Kasten, Ira Papick, Nathalie Sinclair, Chris Hirsch, Bob Reys, Gwen Lloyd, and Mary Ann Huntley, responsible for the selection of speakers and the program;
- the local arrangements staff—directed by Carol Siegel—responsible for the selection of the Field Museum as our first venue and for the wonderful refreshments;
- the speakers—who have given us great ideas and inspiration; and
- you, the participants—without whom there is no conference.

And also we must thank the National Science Foundation (NSF), whose funds supported about two-thirds of this conference.

This is the sixth international conference in mathematics education dealing with curriculum that we have hosted at this university in the past 25 years. Four of these conferences have been under the auspices of the University of Chicago School Mathematics Project and the last two were

Future Curricular Trends in School Algebra and Geometry: Proceedings of a Conference. pages 305–312

run by the Center for the Study of Mathematics Curriculum. Having become emeritus at the university this past January, I think I can say with some confidence that this will be the last conference in which I have a guiding hand. In these closing remarks, I am not going to try to summarize what has happened at this conference; nor will I try to provide a perspective to match the perspectives given by Jim Mamer, Diane Briars, Doug Clements, Jim Fey, and just now by Bill McCallum and Joan Ferrini-Mundy. I hope that you will indulge me if I use this occasion to make some personal remarks.

When I entered the University of Illinois as a freshman in 1959, I knew I wanted to teach mathematics. In the spring of my freshman year, I took my first course in mathematics education. It was basically a course in the new math, a course in the curriculum work of the University of Illinois Committee on School Mathematics. We learned about the distinction between *number* and *numeral*, that is, between the concept and what was written down. We learned about quantifiers such as *for all* and *there exists*. We learned about using properties such as commutativity and associativity and identities and the deduction of the rules for operations with positive and negative numbers. But the fundamental idea that we learned was that all the reasonably isolated skills we had learned in our high school courses could be viewed as resulting from field properties of the real numbers, and later the complex numbers.

We became imbued with the structure of mathematics and became disciples of it. We knew that we were doing mathematics that had been identified later than other mathematics. We were modernizing the curriculum with properties, logic, matrices and other algebraic concepts that had first appeared in the 19th century. We felt good that we were modernizing the curriculum.

In this university course, I was struck by the fact that there were fundamentally different ways to approach the same mathematical ideas. I wanted to know more about alternate ways to develop mathematics. I went to Harvard for my master's to study with Ed Moise, who had developed the SMSG Geometry. From the time I started teaching, I looked at different ways to approach the subject. I went to Michigan for my doctorate because they would allow me to do a dissertation that involved curriculum. There I fell into the use of transformations in learning geometry.

I say I "fell" into transformations. That isn't exactly true. I was searching for a topic for my dissertation and I asked Joe Payne for some ideas and he said, "There are some people interested in using transformations in geometry. You might want to look into that." And so I did and I became enthused. The way that transformations enabled congruence, similarity, and symmetry to be developed was so elegant. And they applied to all figures, not just triangles and other polygons or circles and the common 3-D figures. My first motivation was mathematical. But something happened as Art Coxford and I were developing the geometry course through transformations and teaching it to 10th grade students. We realized that our view of geometry was changing. We saw things in geometric figures we never saw before—properties were due to symmetry rather than to SAS, or SSS. For instance, the opposite sides of a rectangle were congruent not because we could form congruent triangles with the diagonal but because of the reflection symmetry of the figure. A rhombus was a special kind of kite—a figure that we had not seen in U.S. schoolbooks.

Having transformations in geometry not only affected our view of geometry but also our view of algebra. Graphs of functions became geometric figures, with the graph of the sine congruent to the graph of the cosine. All parabolas were similar! The cosine and sine functions could be defined on the unit circle as coordinates of the image of the point (1,0) under rotations. Matrices became important. So did groups. I was awestruck by the relationships between geometry and algebra, and how geometry now had so many applications in the later algebra courses.

It happens that we hit transformations at just the right time, and the work we did with transformations had some influence on other materials. I felt lucky to have hit this at the right time.

Shortly after this work with transformations and matrices and groups in both geometry and second-year algebra, I was encouraged by people to think about first-year algebra. Max Bell was pushing applications throughout mathematics, and I took his lead and developed a first-year algebra course through applications, including probability and statistics. Again it seems that I hit the field with the right idea at just the right time, and that work too had its influence, not just on algebra but on other courses. It certainly influenced the University of Chicago School Mathematics Project (UCSMP), which came some years later. And we felt the UCSMP curriculum, developed before the National Council of Teachers of Mathematics (NCTM) curriculum standards, influenced those standards.

About five years after this algebra course through applications was developed, about 25 years ago, UCSMP began. From the very beginning, we were committed to using the latest in technology. I spoke about computer algebra systems (CAS) in a major talk at the NCTM annual meeting in 1983; I thought CAS would start appearing in school curricula in a couple of years. It has taken longer. Only now are we seeing school curricula appear with a significant amount of CAS work.

Through the years many people told me that they felt I had some sort of knack for foreseeing what happens in curriculum. Early in my career I thought it was luck. Later in my career I started wondering whether the people were right—maybe I had some sort of special knack for latching onto ideas that later became used.

But now, looking back at an entire career, I can see that it was neither luck nor some special knack. I was in the *field* of mathematics curriculum and at the places where cutting-edge work was being done or being considered. In this field, many others were doing the same thing and had been working long before me. Transformations had been used in Europe ever since Felix Klein almost 100 years before Art and I used them in the U.S., and Zoltan Dienes in Hungary and Gustave Choquet in France were talking about the use of transformations with students before we were working with them. Howard Fehr and Jim Fey were working simultaneously with transformations at Columbia University with the Secondary School Mathematics Curriculum Improvement Study. Paul Kelly and Norm Ladd had written a geometry book with a long chapter on transformations.

In applications of mathematics, Max Bell and Henry Pollak and others had been working in applications for years before I got into that arena, and Sol Garfunkel went into it with a vengeance with at the Consortium for Mathematics and its Applications (COMAP). Jim Fey and Kathy Heid were working with CAS. Many others have been involved in the notion that you could approach mathematics differently and change how students view the subject, including Chris Hirsch and Glenda Lappan and Bob and Barbara Reys and many others who have been working in the field for over a quarter of a century, and so many others who joined us in the 1990s with the NSF-supported projects.

Through all these years, the University of Chicago fortunately provided an environment that supported this curriculum work. Max Bell and I were able to use our work to obtain tenure and promotions to full professors at a time when comparable institutions—Harvard, Yale, Princeton, and Stanford—had no mathematics education at all.

In the early 90s, NSF played an important role in the field of curriculum development. By funding over a dozen multi-year curriculum projects, and by encouraging—well, actually, forcing—the people from those projects to get together in what were known as the Gateways Project conferences, NSF caused universities to realize that the study of mathematics curriculum was a viable research area. By funding full curricula, it forced the developers to have to deal with the complexities of the mathematics curriculum and the intersections of curriculum with policy, schools, classrooms, and students. By establishing the Arc, Show-Me, and Compass Centers to get the word out about the curricula, NSF may have been a little over-enthusiastic about trumpeting its own curricula, but it again helped to bring the field together by forcing everyone to realize that curriculum development is a serious field of study driven by what is best for students and teachers in various places and by what will most likely improve mathematics education.

Today, however, we have seen disturbing developments. Despite the very public statements of the U.S. Department of Education of the importance of well-constructed comparative curriculum research, there does not seem to be much of an effort to support the basic curriculum development work that leads to that research. Indeed, sometimes people who use different approaches to school mathematics, we who do curriculum development, are looked at as if we are in alternative medicine, doing something unsanctioned that will surely ruin the nation's students. The notion that *new* curricula must pass some litmus test before they can be used, but older curricula that have failed do not have to pass any test, is a mechanism to quash any sort of curriculum change.

We cannot wait for basic research to underlie everything we do in education. If we did, we would have nothing to teach. For instance, there is virtually no research in the learning of statistics but students must learn statistics. And the gold standard has a moral aspect with respect to technology. Medical studies are often stopped in the middle when it becomes clear that the new method is much better than the old. This is the case with powerful technology. It is not fair to give one group technology and not give the same technology to another group. Similarly, we cannot have research comparing students who are taught statistics (or any other new topic) with those who are not, because it is not fair to give a test to students with content they have not studied. We have to do the best we can, with curricula that are comfortable for students and teachers, and curricula that show that students can learn these ideas. Fortunately, the choice of what to teach is a matter of belief rather than research. The recent report on school algebra of the National Mathematics Advisory Panel makes that patently clear.

That panel did not understand many things. But fundamentally, what they did not realize is that the mathematics curriculum is a living organism that moves in reaction both to its heredity and its environment. Of its heredity, one parent might be said to be pure mathematics and statistics and computer science—and the other parent is applied mathematics—consumer mathematics and quantitative literacy and the nonacademic uses of mathematics in business and everyday life. The environmental influence on mathematics curriculum is the students and the teachers and the schools and the communities in which learning takes place. Those of us who work in curriculum naturally look at what is going on and try to nourish this organism as we think best. In trying to create the healthiest possible organism, we must consider both heredity and environmental factors. It is natural for many of us to be thinking somewhat alike but not exactly alike, just as medical researchers looking at a disease may have similar but different plans of attack. We usually see parts of the organism that need help and other parts that are best left alone.

The two notions—that the mathematics curriculum is a living organism in constant need of examination for its wellness and sickness, and that the study and work of that organism is a field, not just people's hobbies or happenstances of current events—can be seen as *the* construct underlying the existence of the Center for the Study of Mathematics Curriculum (CSMC). This is why the CSMC's work is critical for the future of mathematics education in our country, why we feel it is so important to develop students whose doctoral study involves concentrated work in mathematics curriculum.

It is the total disdain for this work that underlies what disturbs me and some others of us so much in the report of the National Mathematics Advisory Panel on school algebra and—perhaps less so, but for me, just as significant—also about the NCTM Focal Points.

The panel's view of algebra goes up only one parent of the hereditary tree of mathematics, and only one grandparent, and perhaps only one great-grandparent. As I quipped in one of the concurrent sessions (at this conference), it should not have been called the National Mathematics Advisory Panel, but the National Paper-and-Pencil Manipulative Algebraic Skills Advisory Panel. Then its report would make sense. But as is, the report does not make any sense, for it completely ignores all of the other heredity of algebra. And it also completely ignores the change in environment and everything that has been done regarding the algebra curriculum in the past 40 years because of that change in environment.

Regarding the NCTM Focal Points, returning to the living organism analogy, we know everyone should eat fruits and vegetables, but do not know enough to know whether having particular fruits at particular times is better than others. The Focal Points are simply too specific a diet. They constrain rather than expand the repertoire of concepts and approaches that are in some sense sanctioned by specifying too arbitrarily the age at which certain ideas should be encountered. The work in this conference shows us that the age at which serious work in algebra should begin is not at all settled. The fact that various states have differed on when they should introduce or master many topics in arithmetic or geometry is evidence that setting specific grade levels for many topics is rather arbitrary. Furthermore, by doing so one introduces serious gaps. For instance, there is no geometry in grade 6 in the Focal Points. Not in the Focal Points or in the accompanying ideas. None at all. Is there something about that age—ages 11 and 12 for most students—that would indicate geometry should not be studied? Of course not. The NCTM Standards of 1989 and Principles and Standards for School Mathematics (PSSM) of 2000 were much more like broad guidelines (thank you Joan Ferrini-Mundy), with the appropriate balance of latitude and specificity. Let us hope that NCTM steps back in time and realizes the errors of its recent ways. As Diane Resek suggested yesterday in a session, the 1989 Curriculum Standards may be the best document we have created so far.

This conference was titled "Future Curricular Trends in School Algebra and Geometry". Did we pick the right trends? Frankly, we picked safe trends, because the four ideas that we chose for discussion at this conference are not new ideas. Algebra used to be a college subject. The vestige of this is in the fact that there is a course in community colleges called "college algebra" that has the same content that we have in advanced algebra and precalculus mathematics. Statistics used to be taught first in colleges. Geometry used to be absent from early elementary school. Over 40% of 8th graders in the U.S. now are taking an algebra course. Algebra is being taught earlier and earlier, and the work described here is continuing a multi-century trend. It is a safe prediction that what we now call "early algebra" will in the far future not be considered early.

Our second theme, CAS, is relatively new in that computer algebra systems were first developed in the early 1970s. But the notion of developing new algorithms for solving mathematics problems is as old as mathematics itself. The Babylonians and Egyptians described algorithms 3000 years ago; the Chinese developed what we call the Chinese remainder theorem to solve problems that today we call systems of equations in modular arithmetic. In the middle ages there was a conflict between the abacists, those who did arithmetic on the abacus, and the algorithmists, those who used paper-and-writing-implement (there were no pencils at first) algorithms that were new for the day. Newton's method for solving polynomials and the calculus were valuable because they provided powerful algorithms for solving many questions about curves and physics. One use of CAS, to provide easy and automatic ways of solving hitherto very difficult problems, simply follows what has been perhaps the main purpose of mathematics over the years—to solve problems in an efficient way. This, too, is a safe theme. Another use of CAS was not discussed much at this conference—its value as an instructional tool particularly in helping slower students learn algebra. CAS is here to stay and the only question is the speed with which curricula will adapt to handle the new and powerful ways it gives us to look at mathematics and solve problems.

Our third theme, 3-D geometry, is in some sense the most enigmatic of mathematics curriculum themes, but it, too, is nothing new. There is solid geometry in Euclid's *Elements* and trigonometry was first developed from the celestial sphere, not the plane. Our problem has always been to represent the 3-D world on paper. We still have not solved that problem, but new technology shows us that we can represent our 3-D world quite nicely on a 2-D screen by manipulating objects in space. This may be a virtual environment, but isn't paper-and-pencil also a virtual environment? When Hilbert said that mathematics dealt with the marks and symbols drawn on paper, could we not interpret that as saying that mathematics is itself a virtual environment?

The fourth theme, the integration and linking of algebra and geometry, also has ancient origins. The geometric approach to number and what is today's quadratic equations found in Euclid's *Elements*, the analytic geometry of Fermat and Descartes, the calculus of Newton and Leibniz all bring both algebra and geometry to play. Bringing these subjects together is bringing them *back* together after a hiatus of a couple of hundred years while their foundations were being strengthened logically.

Thus it is not much of a prediction to say that the themes of this conference will be significant in keeping the organism healthy in the years to come. Algebra will become a staple of elementary school mathematics, maybe not in the next few years, but surely in the next few centuries. It is as inevitable as arithmetic's moving down over time from college to elementary school. 3-D geometry will come back into the curriculum through technology if for no other reason than the world is still, for almost all practical purposes, 3-dimensional. Geometry and algebra will become more and more intertwined as we see that each represents the other and have the technology to show the simultaneous representations. CAS will revolutionize the learning of algebraic algorithms in the same way that paper-and-pencil technology revolutionized the learning of arithmetic. Some of these things may happen in just a few years; others may take centuries. But they are inevitable, because organisms do not just live, they evolve. We in the Center for the Study of Mathematics Curriculum hope that you have enjoyed looking at these forces moving the living organism of mathematics curriculum.

We hope that you have found the past two days here to be stimulating and informative and wish you safe trips back home. And again, thank you for coming and contributing to the conference.

APPENDIX

SECOND INTERNATIONAL CURRICULUM CONFERENCE

Future Curricular Trends in School Algebra and Geometry
May 2–4, 2008
The University of Chicago and The Field Museum
Chicago, Illinois, USA

FRIDAY, MAY 2

The Field Museum
West Entrance

9:00	Museum opens	
10:30	**Ongoing conference check-in opens**	
10:30–12:30	Light refreshments available	West Lobby

Future Curricular Trends in School Algebra and Geometry: Proceedings of a Conference. pages 313–334

Refreshments are available in the West Lobby during breaks.

12:30–12:45 **Welcome** James Simpson Theater

Zalman Usiskin, The University of Chicago

Barbara Reys, University of Missouri, Columbia

12:45–2:15 **Plenary Session I—***Algebra in Elementary School* James Simpson Theater

Moderator: Kathryn B. Chval, University of Missouri, Columbia

Speakers: Romulo Campos Lins, Universidade Estadual Paulista, Rio Claro, Brazil

Elizabeth Warren, Australian Catholic University, Brisbane

2:15–2:30 Break

2:30–4:00 **Plenary Session II—***Algebra Using Computer Algebra Systems (CAS)* James Simpson Theater

Moderator: Steven W. Ziebarth, Western Michigan University, Kalamazoo

Speakers: Bernhard Kutzler, Austrian Center for Didactics of Computer Algebra, Linz

Kaye Stacey, University of Melbourne, Australia

4:00–4:15 Break

4:15–5:45 **Concurrent Sessions**

Theme I: Early Algebra James Simpson Theater

Moderator: Elizabeth (Betty) Difanis Phillips, Michigan State University, East Lansing

Speakers: Maria Blanton, University of Massachusetts, North Dartmouth

Barbara J. Dougherty, University of Mississippi, University

Comments: Romulo Campos Lins, Universidade Estadual Paulista, Rio Claro, Brazil

Theme II: Algebra Using CAS Lecture Hall 1

Moderator: Mary Ann Huntley, University of Delaware, Newark

Speakers: Al Cuoco, Education Development Center, Newton, MA

M. Kathleen Heid, The Pennsylvania State University, University Park

Comments: Bernhard Kutzler, Austrian Center for Didactics of Computer Algebra, Linz

6:00–8:00 **Reception** West Lobby

6:30–8:00 **Special Exhibit** Holleb Exhibition Gallery

"Mythic Creatures: Dragons, Unicorns and Mermaids" open for private CSMC viewing.

Participants may return to the reception if they wish after viewing the exhibit.

SATURDAY, MAY 3

The University of Chicago
Ida Noyes Hall

8:15–8:45 Continental Breakfast

Meals and refreshments during the breaks are available in the Cloister Club.

8:45–9:00 **Welcome and Announcements** Max Palevsky Cinema

Zalman Usiskin, The University of Chicago

9:00–10:30 **Plenary Session III**—*3-D Geometry* Max Palevsky Cinema

Moderator: Nathalie Sinclair, Simon Fraser University, Burnaby, British Columbia, Canada

Speakers: Claudi Alsina, Universitat Politècnica de Catalunya, Barcelona, Spain

Jean-Marie Laborde, Université Joseph Fourier, Grenoble, France

10:30–11:00 Break

11:00–12:30 **Plenary Session IV**—*Linking Algebra and Geometry* Max Palevsky Cinema

Moderator: Ira J. Papick, University of Missouri, Columbia

Speakers: Keith Jones, University of Southampton, UK

Colette Laborde, Institut Universitaire de Formation des Maîtres, Grenoble, France

12:30–1:30 **Lunch (provided)**

1:30–3:00 **Concurrent Sessions**

Theme I: Early Algebra West Lounge

Moderator: John P. (Jack) Smith III, Michigan State University, East Lansing

Speakers: Maria Blanton, University of Massachusetts, North Dartmouth

Barbara J. Dougherty, University of Mississippi, University

Comments: Elizabeth Warren, Australian Catholic University, Brisbane

Theme III: 3-D Geometry Library

Moderator: Zalman Usiskin, The University of Chicago

Speakers: Thomas F. Banchoff, Brown University, Providence, RI

Michael T. Battista, The Ohio State University, Columbus

Comments: Jean-Marie Laborde, Université Joseph Fourier, Grenoble, France

Theme IV: Integrating and Linking Algebra and Geometry East Lounge

Moderator: Denisse R. Thompson, The University of Chicago and The University of South Florida, Tampa

Speakers: Nicholas Jackiw, KCP Technologies, Emeryville, CA

Diane Resek, San Francisco State University, CA

Comments: Keith Jones, University of Southampton, UK

3:00–3:30 Break

3:30–5:00 **Concurrent Sessions**

Theme II: Algebra Using CAS West Lounge

Moderator: Steven W. Ziebarth, Western Michigan University, Kalamazoo, MI

Speakers: Al Cuoco, Education Development Center, Newton, MA

M. Kathleen Heid, The Pennsylvania State University, University Park

Comments: Kaye Stacey, University of Melbourne, Australia

Theme III: 3-D Geometry Library

Moderator: Robert Reys, University of Missouri, Columbia

Speakers: Thomas F. Banchoff, Brown University, Providence, RI

Michael T. Battista, The Ohio State University, Columbus

Comments: Claudi Alsina, Universitat Politècnica de Catalunya, Barcelona, Spain

Theme IV: Integrating and Linking Algebra and Geometry East Lounge

Moderator: Sarah Kasten, Michigan State University, East Lansing

Speakers: Nicholas Jackiw, KCP Technologies, Emeryville, CA

Diane Resek, San Francisco State University, CA

Comments: Colette Laborde, Institut Universitaire de Formation des Maîtres, Grenoble, France

5:00–5:15 **Announcements** Max Palevsky Cinema

Zalman Usiskin, The University of Chicago

SUNDAY, MAY 4

The University of Chicago
Ida Noyes Hall

8:15–8:45 **Continental Breakfast**

Breakfast and refreshments during the break are available in the Cloister Club.

8:45–9:00 **Announcements** Max Palevsky Cinema

Carol Siegel, The University of Chicago

9:00–10:30	**Panel of Practitioners and Curriculum Developers** Max Palevsky Cinema
	Moderator: Barbara Reys, University of Missouri, Columbia
	Panelists: Diane J. Briars, Pittsburgh Reform in Mathematics Education Project, PA
	Douglas H. Clements, University at Buffalo, NY
	James Fey, University of Maryland, College Park
	Jim Mamer, Rockway Middle School, Springfield, OH
10:30–10:45	Break
10:45–11:45	**Closing Session** Max Palevsky Cinema
	Moderator: Zalman Usiskin, The University of Chicago
	Speakers: Joan Ferrini-Mundy, National Science Foundation, Arlington, VA and Michigan State University, East Lansing
	William G. McCallum, The University of Arizona, Tucson
11:45–12:15	**Closing Remarks** Max Palevsky Cinema
	Zalman Usiskin, The University of Chicago

BIOGRAPHIES OF PRESENTERS

Claudi Alsina is a Professor of Mathematics at the Technical University of Catalonia (UPC). From 2002–04, he served as the General Director of Universities in Catalonia. He was a national delegate to the International Mathematical Union from 1986–97, and a member of the International Program Committees for the 7th, 8th, and 9th International Commissions on Mathematical Instruction. His mathematics education interests include visualization techniques, space geometry, modeling, and applications. Professor Alsina holds the highest Awards for Quality Teaching at the University Level from the UPC and the Catalonian government. An author of 20 books and more than 150 papers, he has given hundreds of talks on popularization and mathematics education in the European Union, North and South America, Japan, Singapore, Australia, and New Zealand.

Thomas F. Banchoff is currently a Royce Family Professor in Teaching Excellence at Brown University. His numerous teaching awards include the National Science Foundation Director's Distinguished Teaching Scholar Award. He has been a Carnegie Foundation Pew Scholar and Carnegie Fellow, and was awarded Doctors of Science, *honoris causa*, from Fairfield University and Rhode Island College. Professor Banchoff received his undergraduate degree from the University of Notre Dame, and his Ph.D. from the University of California, Berkeley. He recently published his 100th article, and his books include *Beyond the Third Dimension, Flatland: The Movie,* and *Linear Algebra Through Geometry* (with John Wermer).

Michael T. Battista is Professor of Mathematics Education in the School of Teaching and Learning at The Ohio State University. He has taught mathematics to students of all ages, from preschool through adult, and has been involved in mathematics teacher education both at the preservice and inservice levels. Most of his research has focused on students' learning of geometry and geometric measurement, and the use of technology in mathematics teaching. He is completing development of a set of cognition-based assessment materials for use in elementary school mathematics and, in his current National Science Foundation grant, for investigating elementary teachers' understanding and use of research-based knowledge about students' mathematical thinking. He is the incoming chairperson of the National Council of Teachers of Mathematics' Research Committee and the incoming editor of the *JRME Monograph Series.*

Maria Blanton is Associate Professor of Mathematics and a Senior Executive Research Associate at the James Kaput Center for Research and Innovation in Mathematics Education at the University of Massachusetts, Dartmouth. Her particular expertise in children's early algebraic thinking and in developing classrooms that foster algebraic thinking has led to numerous presentations and publications. She is co-editor (with James Kaput and David Carraher) of the research volume *Algebra in the Early Grades* and author of *Algebra in the Elementary Classroom: Transforming Thinking, Transforming Practice.* She is currently Principal Investigator on the National Science Foundation project *Understanding Linkages Between Social And Cognitive Aspects Of Students' Transition To Mathematical Proof* which explores teaching and learning proof in undergraduate mathematics. As part of this, she is co-editing a research volume (with Despina Stylianou and Eric Knuth), *Teaching and Learning Proof Across the Grades: A K–16 Perspective.*

Diane J. Briars is a mathematics education consultant, after spending 20 years as Mathematics Director for the Pittsburgh Public Schools. Under her leadership, the Pittsburgh Schools made significant progress in increasing student achievement through standards-based curricula, instruction and assessment. She has served as a member of many national committees, including the National Commission on Mathematics and Science Teaching for the 21st Century, headed by Senator John Glenn, and in leadership roles for various national organizations, including the National Council of Teachers of Mathematics, The College Board, and the National Science Foundation. She is currently President-Elect of the National Council of Supervisors of Mathematics. Ms. Briars earned a Ph.D. in Mathematics Education, a M.S. and B.S. in Mathematics from Northwestern University, and did post-doctoral study in the Psychology Department of Carnegie-Mellon University. She began her career as a secondary mathematics teacher.

Kathryn B. Chval is an Assistant Professor and Co-Director of the Missouri Center for Mathematics and Science Teacher Education at the University of Missouri, Columbia. Professor Chval is also a Co-Principal Investigator for the Center for the Study of Mathematics Curriculum and the Researching Science and Mathematics Teacher Learning in Alternative Certification Models Project, both funded by the National Science Foundation. Prior to joining the University of Missouri, Professor Chval was the Acting Section Head for the Teacher Professional Continuum Program in the Division of Elementary, Secondary and Informal Science Division at the National Science Foundation. She also spent fourteen years at the University of Illinois at Chicago directing National Science Foundation-funded projects. Professor Chval's research interests include effective preparation models and support structures for teachers across the professional continuum, effective elementary teaching of underserved populations, especially English-language learners, and curriculum standards and policies.

Douglas H. Clements is Professor of Education at University at Buffalo, SUNY, where he was granted the Chancellor's Award for Excellence in Scholarship and Creative Activities. He was a member of President Bush's National Mathematics Advisory Panel and is on the National Academies of Sciences/National Research Council Committee on Early Childhood Mathematics. His primary research interests lie in the areas of the learning and teaching of geometry, computer applications in mathematics education, and the early development of mathematical ideas. He has published over 90 refereed research studies, 6 books, 50 chapters, and 250 additional publications. Currently, Professor Clements is Principal Investigator on a research project, *Scaling Up TRIAD: Teaching Early Mathematics for Understanding with Trajectories and Technologies* funded by the U.S. Department of Education's Institute of Education Sciences (IES). Professor Clements has directed or co-directed over 15 other projects.

Al Cuoco is a Senior Scientist and Director of the Center for Mathematics Education at the Education Development Center (EDC). A student of Ralph Greenberg, he received a Ph.D. in mathematics from Brandeis, with a thesis in algebraic number theory. Mr. Cuoco taught high school mathematics to a wide range of students in the Woburn, Massachusetts, public schools from 1969–1993. At EDC, he has worked in curriculum development, professional development, and education policy. He currently co-directs a high school curriculum development project, a project to develop a high school linear algebra curriculum, and several professional development projects (in collaboration with the mathematics department at Boston University and the Institute for Advanced Study). His favorite publication is his 1991 article in the *American Mathematical Monthly*, "Visualizing the *p*-adic Integers," described by his wife as "an attempt to explain a number system no one understands with a picture no one can see."

Barbara J. Dougherty is a Professor of Mathematics Education and Director of the Center for Educational Research and Evaluation at The University of Mississippi. She holds a Ph.D. from the University of Missouri-Columbia. With experience in grades 1–12 in mathematics and special education, she has authored curriculum materials and conducted research studies related to improving student learning. She is a past member and chair of the National Council of Teachers of Mathematics Research Committee and is a member of the Essential Understandings for Teaching and Learning Editorial Panel. She is the director of an elementary mathematics research project (LeAD ME), of a Mathematics-Science Partnership Grant (Project DELTA2), and of a secondary algebra and technology project funded by the National Science Foundation. She is the editor of the *Research in Mathematics Education* series, published by Information Age Publishing.

Joan Ferrini-Mundy is the Director of the National Science Foundation's (NSF) Division of Research on Learning in Formal and Informal Settings in the Directorate for Education and Human Resources. She is a University Distinguished Professor of Mathematics Education and Assistant Vice President for Science, Technology, Engineering and Mathematics Education Research and Policy at Michigan State University. She has served on the Board of Directors of the National Council of Teachers of Mathematics (NCTM), chaired the Writing Group for NCTM's *Principles and Standards for School Mathematics*, and completed a term as a member of the Board of Governors of the Mathematical Association of America in 2006. Her research interests include calculus teaching and learning, the development and assessment of teachers' mathematical knowledge for teaching, and the improvement of student learning in K–12 mathematics and science. She served on the President's National Mathematics Advisory Panel.

James Fey is Professor Emeritus in the Department of Mathematics and the Department of Curriculum & Instruction at the University of Maryland where he taught content and methods courses for prospective middle and high school mathematics teachers and mathematics education graduate students. He has been an author of the *Connected Mathematics* middle school and *Core-Plus Mathematics* curriculum materials development projects since their inception, and has been especially involved in the design and writing of units in the algebra content strands. His earlier related work on innovative approaches to school algebra focused on development of materials to support problem-based instruction that takes advantage of calculator and computer tools for learning, teaching, and problem solving.

M. Kathleen Heid is Distinguished Professor in Mathematics Education at The Pennsylvania State University. She has a B.A. in mathematics from The Catholic University of America and an M.A. and Ph.D. in mathematics education from the University of Maryland. She has taught mathematics (middle school, high school, and early college) and mathematics education (undergraduate and graduate). Her research has focused on technology in mathematics teaching and learning (particularly algebra and calculus) and mathematical knowledge of prospective secondary teachers. She has co-authored technology-intensive curricula and co-edited *Research on Technology and the Teaching and Learning of Mathematics: Syntheses, Cases, and Perspectives.* Having served on the Board of Governors for the Mathematical Association of America, as secretary of the American Educational Research Association Special Interest Group for Research in Mathematics Education, and on the Board of Directors for the National Council of Teachers of Mathematics, she is currently Editor-designate of the *Journal for Research in Mathematics Education.*

Mary Ann Huntley is an Assistant Professor in the Department of Mathematical Sciences with a joint appointment in the School of Education at the University of Delaware. She was awarded a National Academy of Education/Spencer Postdoctoral Fellowship (2003) and the American Association of Colleges for Teacher Education's Outstanding Dissertation Award (1997). Building on her experience examining and analyzing mathematics classroom practice at all grade levels in schools across the U.S., her primary research interest involves investigating the relationships between mathematics curricula, teaching, and students' learning, especially at the middle- and high-school levels. Prior to her current position she was a program manager at the National Science Foundation, and before that conducted research, evaluation, and curriculum analysis on numerous national and local projects. She earned her Ph.D. at the University of Maryland, College Park. She has a B.A. in mathematics and computer science, and an M.S.

in applied mathematics. She worked as an applied mathematician for five years.

Nicholas Jackiw is the Chief Technology Officer of KCP Technologies, and is also the software designer responsible for *The Geometer's Sketchpad®*. As one of the founding members of the Visual Geometry Project at Swarthmore College, he was responsible for the design and development of all of the VGP interactive software. He directed *Sketchpad* product development at Key Curriculum Press from 1990–98 when, with others, he developed Key Curriculum Press' software department into KCP Technologies. In addition to designing software, he is the chief programmer of several incarnations of *Sketchpad.* He works with schools in conducting field-testing and software evaluation. He has been principal investigator and senior scientist on Small Business Innovative Research projects investigating dynamic geometry's impact and potential, and has written numerous articles on the subject. He is also currently an adjunct professor at Simon Fraser University, Burnaby, British Columbia, Canada.

Keith Jones is an Associate Professor in Pedagogy and Curriculum at the University of Southampton, United Kingdom, where he leads the Mathematics Education Research Group. His interests in mathematics education span geometrical thinking and reasoning, the use of technology, and teacher education and professional development. He instigated the Geometry Working Group of the British Society for Research into Learning Mathematics and is on the editorial board of the society's official journal, *Research in Mathematics Education.* He founded the Special Interest Group on Mathematics Education within the British Educational Research Association. In 2000–01, he co-edited a special triple-issue of the journal *Educational Studies in Mathematics* on aspects of mathematical proof and proving while using dynamic geometry software. During 2006, he co-edited four special issues of the *International Journal of Technology in Mathematics Education*, including issues focusing on geometry and on algebra. He is a founding member of the thematic group on Tools and Technologies in Mathematical Didactics of the European Society for Research in Mathematics Education and led the group from 2000–03.

Sarah Kasten is currently a doctoral student at Michigan State University studying teacher education. She earned a master's degree from The Ohio State University in mathematics education and was a high school mathematics teacher for three years.

Bernhard Kutzler is vice-president of the Austrian Center for Didactics of Computer Algebra and a lecturer at the University of Linz in Austria.

He holds a Ph.D. in mathematics with specialization in computer algebra. Since 1991, his research interest has been the integration of computer algebra into mathematics education. He is the author of numerous articles and books and the organizer of various international conferences in this area.

Colette Laborde is Professor Emerita at the Joseph Fourier University, Grenoble, France. Her research deals with the integration of computers in the teaching and learning of geometry. She is involved in the project Cabri-géomètre, which creates dynamic geometry software programs for plane geometry and 3-D geometry. From 1985–2003, she was head of the doctoral program in Mathematics and Science Education at the Joseph Fourier University. From 2003–06, she led a research team on technology in mathematics education at this university. She was a member of the executive committee of the International Commission for Mathematics Instruction (ICMI) from 1994–98. She is the co-chair of the Topic Study Group on "New Technologies in the Learning and Teaching of Mathematics" for ICME11 in July 2008. She has edited several books about mathematics education and is a member of several editorial boards of international journals in mathematics education.

Jean-Marie Laborde graduated in mathematics from the École Normale Supérieure and began working in the Laboratory for Computer Sciences and Applied Mathematics at Grenoble University (now the Joseph Fourier University), France. He received his Docteur d'Etat in Mathematics in 1977. He has researched the use of geometric methods for the study of different classes of graphs, especially hypercubes. His interests also included automatic theorem proving. In 1981 he and a group of French scholars started the Cabri project and in 1982 he founded the Laboratory for Discrete Mathematics and Research in Mathematics Education at Grenoble University. He taught Mathematics and Computer Sciences and has been appointed a university professor in France and Germany. In 1987, a number of students and young researchers joined the Cabri project to start Cabri-géomètre, a sketchpad for geometry. He later became Research Director at the French National Center for Scientific Research (CNRS) and the head of the Cabri-geometry Project, a collaborative IMAG-Texas Instruments project, involving more than 25 people at the Joseph Fourier University. In 2000, Professor Laborde founded the start-up company Cabrilog. From June 2000 to June 2004 he worked for CNRS at Cabrilog and since then he has been devoted to Cabrilog and the development of Cabri-Technology. Professor Laborde is the author of more than 80 scientific papers.

Romulo Campos Lins is an Associate Professor in the Mathematics Department and a member of the Postgraduate Program on Mathematics Education, at São Paulo State University, Rio Claro, Brazil. His research interest is meaning production and the theory of knowledge as related mainly to mathematics education, in particular early algebra education and the role of "advanced" mathematics courses (e.g., calculus, analysis, and linear algebra) in the preparation of future mathematics teachers. Professor Lins also coordinates a project in collaboration with the Ministry of Education that works with the mathematical development of primary school teachers in different Brazilian states.

Jim Mamer has been a middle school math teacher for his 20-year career. For the past 10 years he has implemented the *Connected Mathematics* curriculum with his students. Mr. Mamer has received extensive training from the authors of the *Connected Mathematics* Project and has trained many teachers across the country to implement the curricula. He currently teaches middle school students as a "math coach" for the Clark-Shawnee Local School District in Springfield, Ohio. Mr. Mamer has a degree in elementary education and a master's degree in supervision from Wright State University. He is National Board Certified in early adolescence/young adulthood mathematics and has won numerous awards including the Presidential Award for Excellence in Teaching.

William G. McCallum is a University Distinguished Professor of Mathematics and Director of the Institute for Mathematics and Education at the University of Arizona. He received his Ph.D. in Mathematics from Harvard University. After spending two years at the University of California, Berkeley, and one at the Mathematical Sciences Research Institute in Berkeley, he joined the faculty at the University of Arizona. In 1989, he joined the Harvard calculus consortium, and is the lead author of the consortium's multivariable calculus and college algebra texts. He spent 1993–94 at the Institut des Hautes Etudes Scientifiques, and 1995–96 at the Institute for Advanced Study on a Centennial Fellowship from the American Mathematical Society. In 2005, he received the Director's Award for Distinguished Teaching Scholars from the National Science Foundation. His professional interests include arithmetical algebraic geometry and mathematics education. He has received grants and written articles, essays, and books in both areas.

Ira J. Papick is a Professor of Mathematics and a Curators' Teaching Professor at the University of Missouri, Columbia. He has bachelor's and master's degrees in Mathematics Education and a Ph.D. in Mathematics from Rutgers University. His research areas are in commutative algebra and mathematics education. He has published research papers in both areas, co-authored the research level book, *Prufer Domains*, and authored the undergraduate textbook for middle grade mathematics teachers, *Algebra Connections*. Professor Papick has extensive involvement in K–12 mathematics education with particular emphasis on teacher preparation. He is the recipient of five University of Missouri teaching awards, the most prominent of which is the Intercampus Presidential Award for Outstanding Teaching, 2001.

Elizabeth (Betty) Difanis Phillips is a Senior Academic Specialist in the Division of Mathematics and Science Education at Michigan State University. She has been Chair and Editor of the Michigan Council of Teachers of Mathematics monograph series, a member of the Leading Mathematics-Education into the 21st Century project, and a member of the National Council of Teachers of Mathematics (NCTM) Addenda project, co-authoring the book, *Patterns and Functions for Middle Grades Teachers.* She was a Co-Director of the National Science Foundation (NSF)-funded Algebra for the Twenty-First Century conference held in 1992 and chair of the NCTM Algebra Working Group (1994–1995). She has been a member and Chair of the NCTM Nominations and Election Committee. She is the author of numerous papers and books and a co-author of *The Connected Mathematics Project* (CMP). She was also a co-director of a NSF-funded teacher enhancement grant to develop teams of mathematics educators who can assume a leadership role in implementing Standards-based curricula across the country. Her current work is developing professional development materials using classroom videos to support the implementation of CMP. She is a Principal Investigator of the CSMC.

Diane Resek is a Professor Emerita of Mathematics at San Francisco State University. She received her Ph.D. from the University of California, Berkeley in the area of algebraic logic. She has been active in curriculum development for students in middle school, secondary school, and college. She was a co-author of *The Interactive Mathematics Program,* which is one of the comprehensive secondary mathematics programs developed through National Science Foundation (NSF) funding. She has developed mathematics courses for teachers at the elementary, middle, and secondary school levels. Currently she is co-director of REvitalizing ALgebra (REAL), one of the NSF Math and Science Partnership targeted programs.

Barbara Reys is the Lois Knowles Professor of Mathematics Education and Director of the Center for the Study of Mathematics Curriculum at the University of Missouri, Columbia. She has an undergraduate degree in mathematics and a Ph.D. in mathematics education. She served as a writing group leader for the National Council of Teachers of Mathematics' (NCTM) *Principles and Standards for School Mathematics* and on the NCTM Board of Directors. She is the President-Elect of the Association of Mathematics Teacher Educators. Her current research focuses on the role and influence of curriculum standards and textbooks in teaching and learning mathematics, K–12. She recently published an analysis of mathematics curriculum standards produced by 42 state departments of education, documenting the lack of consensus regarding grade placement of important topics in K–8 mathematics.

Robert Reys is a Curators' Professor of Mathematics Education at the University of Missouri. He has done research in estimation, mental computation, and number sense. He has authored/co-authored over 180 articles in professional journals and 30 books. Currently he is one of the co-principal investigators of the CSMC.

Nathalie Sinclair studied mathematics as an undergraduate at McGill University in Quebec and received a master's degree in mathematics from Simon Fraser University in British Columbia where, at the Centre for Experimental and Constructive Mathematics, she built interactive visualization software for mathematicians and students. She received a doctorate in mathematics education at Queen's University, where her dissertation examined the important roles that aesthetics play in the development and communication of mathematics. Her first academic position was at Michigan State University, where she published *The History of the Geometry Curriculum in the United States* under the auspices of the Center for the Study of Mathematics Curriculum, *Mathematics and Beauty: Aesthetic Approaches to Teaching Children*, and co-edited *Mathematics and the Aesthetic: New Approaches to an Ancient Affinity*. She is now an assistant professor at Simon Fraser University, where she is investigating the pedagogical possibilities of 3-D immersive environments such as Second Life.

John P. (Jack) Smith III is an educational psychologist by training and an analyst of human learning and development, particularly in the domain of mathematics. He is currently Associate Professor in the Department of Counseling, Educational Psychology, and Special Education, in the College of Education at Michigan State University (MSU). Some of his recent research has concerned middle school and high school students' reactions to and transitions between "reform" and "traditional" curriculum; the origins of students' struggle with spatial measurement; and the structure and content of states' K–8 mathematics standards in geometry and measurement. His teaching focuses on psychological perspectives and empirical research on learning school subject matter. He holds bachelor's degrees in mathematics and philosophy and master's and Ph.D. degrees from the Graduate School of Education at the University of California, Berkeley. He is currently the Director of the Educational Psychology and Educational Technology doctoral program and a member of the Mathematics Education Faculty Group at MSU.

Kaye Stacey is the Foundation Professor of Mathematics Education at the University of Melbourne and leader of the science and mathematics education group. She works as a researcher and teacher educator, in undergraduate and post-graduate courses, training teachers for both primary and secondary schools. She has written many practical books and articles for teachers as well as many research articles. Professor Stacey's research interests center on mathematical problem solving and the mathematics curriculum, particularly in the algebra and number strands and in the challenges that are faced in adapting to the new technological environment. She has received an Australian government medal for her work with Australian national and state curriculum projects, and as the mathematics education expert on the Australian advisory committee for PISA and TIMSS. She was educated at the University of New South Wales, and obtained her Ph.D. in number theory at Oxford University.

Denisse R. Thompson is a Professor of Mathematics Education at the University of South Florida (USF), Tampa, where she teaches undergraduate and graduate mathematics methods courses for K–12 teachers. These courses include reading issues in mathematics, history, and research related to mathematics education curriculum, and assessment issues in mathematics education. With Sharon Senk, she has coauthored a book which documents research on many contemporary mathematics curriculum projects, *Standards-Based School Mathematics Curricula: What Are They? What Do Students Learn?* She has co-authored a mathematics methods text for middle-grades teachers and recently co-authored a book on Mathematical Literacy. Professor Thompson became interested in curriculum development through her work with the University of Chicago School Mathematics Project (UCSMP), first as an author on *UCSMP Advanced Algebra* and then as an editor on *Precalculus and Discrete Mathematics* while completing her doctoral studies. Since 2005, she has been the Director of Evaluation for the Third Edition of the UCSMP Secondary Component with responsibility for directing the field studies as the third edition materials have been developed.

Zalman Usiskin is a Professor Emeritus of Education at the University of Chicago, where he was a faculty member from 1969–2007. His curriculum research has included the incorporation of geometric transformations, related ideas from linear and abstract algebra, and applications and mathematical modeling into the curricula of average students. He has been associated with the University of Chicago School Mathematics Project (UCSMP)and overseen the development of its curriculum for middle school and high school since UCSMP's inception in 1983 and, since 1987, he has been its overall director. His most recent book, co-authored with three others, is *The Classification of Quadrilaterals: A Study of Definition.* He has been a member of the Mathematical Sciences Education Board, the Board of Directors of the National Council of Teachers of Mathematics, the mathematics standing committee of the National Assessment of Educational Progress, and member and chair of the United States National Commission on Mathematics Instruction. He is a co-Principal Investigator of the CSMC.

Elizabeth Warren is Associate Dean of Research at the Australian Catholic University. She has a deep interest in mathematics education, with experiences ranging from secondary mathematics teaching to working with preschool children, supporting them in exploring the "big ideas" of mathematics in a play-based environment. The beliefs that drive her continual passion for mathematics learning are (a) all students have a right to engage with mathematics, and (b) this engagement is dependent on the types of activities utilized in the classroom and the discussions that ensue. She has published extensively in research journals and also produced a series of books for elementary classroom teachers called *Algebra for All.*

Steven W. Ziebarth is an Assistant Professor of Mathematics Education in the Department of Mathematics at Western Michigan University. Since 1992, he has been the evaluation coordinator for three phases of the National Science Foundation (NSF)-sponsored Core-Plus Mathematics Project including a five-year longitudinal study and serving as principal evaluator for the second-edition of *Core-Plus Mathematics.* His evaluation experience also includes being lead evaluator on other NSF Projects including the Project to Implement the NCTM Standard in Discrete Mathematics, the Iowa PRIME-Team Local Systemic Change Initiative, the Technology and Reform of the Undergraduate Preparation of Elementary/Middle School Teachers (TRUMPET), and the Mathematics Online Support for Teachers (MOST) Project (jointly with the North Central Regional Educational Laboratory). He has also performed evaluation work at the state level through Eisenhower-funded projects and special evaluation tasks for local school districts and university program reviews.

PARTICIPANTS

Allen, Richard; St. Olaf College
Anderson, Rick; Eastern Illinois University
Aquino, Nadja; University of Chicago Laboratory Schools, Illinois
Baker, Marvin; DaVinci Academy, Utah
Barbour, Robin; North Carolina Department of Public Instruction

Barry, Dorothy; Johns Hopkins University
Bonds, Chulrae; Miller Grove High School, Georgia
Boyd, Consuela; Miller Grove High School, Georgia
Brothers, Gosia; Texas Instruments
Brown, Heidi; Greensburg Salem School District, Pennsylvania
Browning, Christine; Western Michigan University
Campbell, Christine; Villa Duchesne High School, Missouri
Caraco, Michael; Burr and Burton Academy, Vermont
Chavez, Oscar; University of Missouri-Columbia
Clyne, Amy; Waukegan High School, Illinois
Conner, Kimberly; Vanderbilt University
Converse, Timothy; Duquesne University
Cox, Chris; Kalamazoo Public Schools, Michigan
Cox, Dana; Western Michigan University
Cummings, Liza; University of Missouri-Columbia
Davis, Carmen; Boston College
Dietiker, Leslie; Michigan State University
Doran, Erik; Greensburg Salem School District, Pennsylvania
Foley, Gregory; Ohio Univeristy
Fry, Mark; Texas Instruments
Gartner, Phil; Glenbrook South High School, Illinois
Goetting, Mary; Concordia University Chicago
Goyette, Dennis; Talent Development High School, Johns Hopkins University, Maryland
Harper, Susan; McGraw-Hill
Hatcher, Alan; Dallas Independent School District, Texas
Hirsch, Christian; Western Michigan University
Hopkins, Bill; University of Texas
Horvath, Aladar; Michigan State University
Hsiao, Joy; High School for Arts & Business, New York
Hurtado-Fuentes, Luis; Math Department Dallas Independent School District, Texas
Isaacs, Andy; University of Chicago
Jackson, Christa; University of Missouri-Columbia
Jahnke, Anette; National Center for Mathematics Education, Gothenburg, Sweden
Jayson, Jennifer; George Mason HS, Virginia
Johnson, Eddie; Miller Grove HS, Georgia
Jones, Danny; John Hopkins University
Kaduk, Cathy; Naperville School District, Illinois
Kania, Irena; Collins High School, Illinois
Kean, Lesa; Illinois Institute of Technology
Kish, Denise; Harbor Beach Community School, Michigan
Koebley, Sarah; Hudson City School District, Ohio
Krouse, Janice; Illinois Math & Science Academy, Illinois
Kujovsky, Matt; Burr and Burton Academy, Vermont
Lanie, Nicole; Western Michigan University
Lappan, Glenda; Michigan State University
Lenaghan, Aaron; Community Consolidated School District 62, Illinois

Lenar, Linda; Harbor Beach Community School, Michigan
Lerner, Mary Elizabeth; Charlotte Catholic High School, North Carolina
Lovanio, Marlene; Connecticut State Department of Education
Mackrell, Kate; Kingston, Ontario, Canada
Madden, Sandra; Western Michigan University
Males, Lorraine; Michigan State University
Manley, Mary Rita; Burr and Burton Academy, Vermont
Martinez, Bertha A.; Lake Geneva Middle School, Wisconsin
Masters, Robin; Francis Parker School, Illinois
Matsko, Vince; Illinois Mathematics and Science Academy, Illinois
McConnell, John; ECRA Group
McNaught, Melissa; University of Missouri-Columbia
Miceli, Kathy; Hudson High School, Ohio
Montague, Angela; Charlotte Catholic High School, North Carolina
Moore, Diane; Western Michigan University
Mortensen, Erika; University of Chicago Charter School, Illinois
Mowry, Cindy; Burr and Burton Academy, Vermont
Nevels, Nevels; University of Missouri-Columbia
Nolan, Maureen; Naperville School District 203, Illinois
O'Connell, Vince; Texas Instruments
Padecky, Laurie; Community Consolidated School District 93, Illinois
Pennisi, Jean; Ridgewood High School, Illinois
Picard, Patricia; Hudson City School District, Ohio
Pomerenke, Sarah; University of Missouri-Columbia
Popovic, Gorjana; Illinois Institute of Technology
Randolph, Jernita; Miller Grove High School, Georgia
Rasmussen, Steven; Key Curriculum Press
Rathouz, Maggie; University of Michigan
Ritsema, Beth; Western Michigan University
Rodriguez, Josephine; Pearson Education
Rogers, Kevin; Miller Grove High School, Georgia
Ross, Daniel; University of Missouri-Columbia
Rubenstein, Rheta; University of Michigan
Russell, Craig; University of Illinois Laboratory Schools, Illinois
Russell, Mary; North Carolina Department of Public Instruction
Sanchez, Stej; Dallas Independent School District, Texas
Sanford, Alison; Villa Duchesne School, Missouri
Schneider, Kim; Jewish Community High School, California
Sherrod, Maggie; Texas Instruments
Sherwin, Katherine; New Canaan Country School, Conneticut
Shields, Alvin; Hunter College Campus Schools, New York
Shihadeh, Eileen; Texas Instruments
Shin, Soo Yeon; Purdue University
Siegel, Charles; Chicago, Illinois
Spohn, David; Hudson High School, Ohio
Stalmack, Richard; Illinois Mathematics and Science Academy, Illinois
Stockstill, Meggen; Harbor Day School, California

Suddreth. Diana; Utah State Office of Education
Switzer, John; University of Missouri-Columbia
Tepavchevich, Louise; Cass Junior High School Distrcit 63, Illinois
Thompson, Patrick; Arizona State University
Threlkeld, John; Graland Country Day School, Colorado
Thurbee, Joshuah; University of Chicago Charter High School, Illinois
Usiskin, Karen; Pearson Education
Vonder Embse, Charles; Central Michigan University
Walker, Diane; Girls Preparatory School, Tennessee
Wang, Sasha; Michigan State University
Warshauer, Hiroko; Texas State University
Wellman, Lydia; Texas Instruments
Whiteley, Walter; York University, Canada
Wickwire, Murray; Horizon Research, Inc.
Williams, Brooke; Schwartz Elementary/Middle School, Oklahoma
Wiltjer, Mary; Oak Park & River Forest High School, Illinois
Winningham, Noreen; University of Chicago Yew, Feiye; Ohio State University
Zimmermann, Gwen; Adlai E. Stevenson High School, Illinois

SPEAKERS

Alsina, Claudi; Universitat Politecnica de Catalunya; ETSAB. Av. Diagonal 649; Barcelona, E08028 Spain; Claudio.alsina@upc.edu

Banchoff, Thomas; Brown Univ.; Mathematics Department; Box 1917; Providence, RI 02912; Thomas_Banchof1@brown.edu

Battista, Michael; Ohio State Univ.; 209 Arps Hall; 1945 N. Arps St.; Columbus, OH 43210; battista.23@osu.edu

Blanton, Maria; Univ. of Massachusetts—Dartmouth; 285 Old Westport Rd.; N. Dartmouth, MA 02747; mblanton@umassd.edu

Briars, Diane; 315 Olde Chapel Trail; Pittsburgh, PA 15238; djbmath@comcast.net

Chval, Kathryn; Univ. of Missouri-Columbia; 120 Townsend Hall; Columbia, MO 65211; chvalkb@missouri.edu

Clements, Douglas; Univ. at Buffalo, SUNY; 212 Baldy Hall; Buffalo, NY 14260; clements@buffalo.edu

Cuoco, Al; Education Development Center; 55 Chapel Street; Newton, MA; acuoco@edc.org

Dougherty, Barbara; Univ. of Mississippi; 1111 Jackson Ave. #8; Oxford, MS 38655; bdougher@olemiss.edu

Ferrini-Mundy, Joan; National Science Foundation; 4201 Wilson Boulevard; Arlington, VA 22230; jferrini@nsf.gov; and Michigan State Univ.; 211 Kedzie Hall; East Lansing, MI 48824; jferrinie@msu.edu

Fey, James; Univ. of Maryland; College Park, MD 20742; jimfey@umd.edu

Grouws, Douglas; Univ. of Missouri-Columbia; 303 Townsend Hall; Columbia, MO 65211; grouwsd@missouri.edu

Heid, M. Kathleen; The Pennsylvania State Univ.; 271 Chambers Building; University Park, PA 16802; mkh2@psu.edu

Huntley, Mary Ann; Univ. of Delaware; Department of Mathematical Sciences; Newark, DE 19716; huntley@math.udel.edu

Jackiw, Nicholas; KCP Technologies; 1150 65th St.; Emeryville, CA 94608; njackiW@keypress.com

Jones, Keith; Univ. of Southampton; Sch. of Education; Southampton, SO 17 IBJ, United Kingdom; d.k.jones@soton.ac.uk

Kasten. Sarah; Michigan State Univ.; A719 Wells Hall; East Lansing, MI 48824; kastens l@msu.edu

Kutzler, Bernhard; Austrian Center for Didactics of Computer Algebra; Hasnerstr. 9/10; Linz 4020, Austria; b.kutzler@aon.at

Laborde, Colette; Joseph Fourier Univ.; Diam-Lig, 46 av Felix Viallet; Grenoble, 38000 France; colette.laborde@imag.fr

Laborde, Jean-Marie; Joseph Fourier Univ.; 6 place Robert Schumann; Grenoble 38000, France; jean-marie.laborde@cabri.com

Lins, Romulo; Universidade EstaduaI Paulista; Department of Mathematics; Av. 24A, 1515; 13506-700 Rio Claro, SP; Brazil; romlins@rc.unesp.br

Mamer, Jim; Clark-Shawnee Local Schs.; 3500 West National Rd.; Springfield, OH 45503; caraandjim@earthlink.net

McCallum, William; Univ. of Arizona; Department of Mathematics; Tucson, AZ 85721; wmc@math.arizona.edu

Papick, Ira; Univ. of Missouri-Columbia; Mathematics Department; Columbia, MO 65211; papicki@missouri.edu

Phillips, Elizabeth; Michigan State Univ.; A717 Wells Hall; East Lansing, MI 48824; ephillips@math.msu.edu

Resek, Diane; San Francisco State Univ.; 1600 Holloway Ave.; San Francisco, CA 94132; resek@sfsu.edu

Reys, Barbara; Univ. of Missouri-Columbia; 303 Townsend Hall; Columbia, MO 65211; reysb@missouri.edu

Reys, Robert; Univ. of Missouri-Columbia; 303 Townsend Hall; Columbia, MO 65211; reysr@missouri.edu

Sinclair, Nathalie; Simon Fraser Univ.; Faculty of Education; Burnaby, BC V5AIS6 Canada; nathsino@sfu.ca

Smith III, John P. (Jack); Michigan State Univ.; 513H Erickson Hall; East Lansing, MI 48895; jsmith@msu.edu

Stacey, Kaye; Univ. of Melbourne; Melbourne 3010; Australia; k.stacey@unimelb.edu.au

Thompson, Denisse R.; Univ. of South Florida; EDU 162; 4202 East Fowler Avenue; Tampa, FL 33620; thompson@tempest.coedu.usf.edu; and Univ. of Chicago; 6030 S. Ellis Avenue; Chicago, IL 60637; denisse@uchicago.edu

Usiskin, Zalman; Univ. of Chicago; 6030 S. Ellis Ave.; Chicago, IL 60637; z-usiskin@uchicago.edu

Warren, Elizabeth; Australian Catholic University; P.O. Box 456; Virginia QLD 4014; Australia; elizabeth.warren@acu.edu.au

Ziebarth, Steven; Western Michigan Univ.; 4429 Everett Tower; Kalamazoo, MI 49008; steven.ziebarth@wmich.edu

CONFERENCE PLANNING COMMITTEE

Zalman Usiskin (Chair)
Christian Hirsch
Mary Ann Huntley
Sarah Kasten
Gwen Lloyd
Ira Papick
Robert Reys
Nathalie Sinclair

LOCAL ARRANGEMENTS

Carol Siegel (Chair); cssiegel@uchicago.edu
Kathleen Andersen
Catherine Ballway
Brandon Bourgeois
Daniel Boutwell
Allison Burlock
Carlos Encalada
Meri Fohran; mbfohran@uchicago.edu
Clare Froemel
Ernie Froemel
Isaac Greenspan
Natalie Jakucyn
Matt McCrea
Emily Mokros
Scott Neff
Julian Owens
Jennifer Perton
Katie Rich
Miriam Rudavsky-Brody
Luke Sandberg
Sarah Schieffer
Gary Spencer
Yayan Zhang

LaVergne, TN USA
23 July 2010
190668LV00002B/2/P

9 781617 350061